Dietary Fibers

Edited by Viduranga Y. Waisundara

Published in London, United Kingdom

IntechOpen

Supporting open minds since 2005

Dietary Fibers
http://dx.doi.org/10.5772/intechopen.95197
Edited by Viduranga Y. Waisundara

Contributors
Jitendra Kumar, Nampher Masharing, Murli Dhar Mitra, Sonia Sangwan, H.B Singh, Kavita Rani, Ranu Agrawal, I. Gusti Ayu Nyoman Danuyanti, Z.S. Ahmad Fahrurrozi, Oluwafunmilayo Dorcas Adegbaju, Gloria Aderonke Otunola, Anthony Jide Afolayan, Ceren Gezer, Gözde Okburan, Viduranga Y. Waisundara, Kanchana Priyadarshani Samarasinghe, Chamodya R. Dharmadasa

First published in London, United Kingdom, 2022 by IntechOpen
IntechOpen is the global imprint of INTECHOPEN LIMITED, registered in England and Wales, registration number: 11086078, 5 Princes Gate Court, London, SW7 2QJ, United Kingdom
Printed in Croatia

British Library Cataloguing-in-Publication Data
A catalogue record for this book is available from the British Library

Additional hard and PDF copies can be obtained from orders@intechopen.com

Dietary Fibers
Edited by Viduranga Y. Waisundara
p. cm.
Print ISBN 978-1-83969-635-0
Online ISBN 978-1-83969-636-7
eBook (PDF) ISBN 978-1-83969-637-4

Meet the editor

Dr. Viduranga Waisundara obtained her Ph.D. in Food Science and Technology from the Department of Chemistry, the National University of Singapore in 2010. She was a lecturer at Temasek Polytechnic (TP), Singapore from July 2009 to March 2013. She relocated to her motherland Sri Lanka and spearheaded the Functional Food Product Development Project at the National Institute of Fundamental Studies from April 2013 to October 2016. She was a senior lecturer on a temporary basis at the Department of Food Technology, Faculty of Technology, Rajarata University of Sri Lanka. She is currently the deputy principal of the Australian College of Business and Technology, Kandy Campus, Kandy, Sri Lanka. She is also the present Global Harmonization Initiative (GHI) ambassadress to Sri Lanka.

Contents

Preface

As a component of food, dietary fibers gained importance when their functional properties came into the limelight. Their role in gut health was identified to be significant, and therefore, they are being incorporated into food products in the market to enhance the health benefits.

This book primarily covers two areas: the fundamentals of dietary fibers and their health benefits. Under the fundamentals of dietary fibers, the chemical, structural, and biochemical aspects of a food component are covered. As for health benefits, emphasis is primarily placed on their benefits imparted to gut microorganisms. The chapters covering these two areas are positioned in such a way that readers who do not come from the relevant academic background can go through the content and realize the importance of dietary fibers for themselves.

As a book that is highly relevant to food chemistry, it is expected that the contents of the chapters will add to the existing knowledge on dietary fibers and also open doors to research that would widen the horizons concerning their importance in health and wellness. From an industrial perspective, it is anticipated that knowledge shared herein would benefit food manufacturers in making scientific judgments about the incorporation and processing of dietary fibers in their food products without causing damage to the functional properties.

I would like to take this opportunity to extend my appreciation to the authors who have contributed all the wonderful and informative chapters to this book. Also, heartfelt appreciation goes to the Publisher IntechOpen with whom I have worked on several publications prior to this one. Last but not least, my appreciation goes to the Publishing Process Managers Ms. Marijana Francetic and Ms. Marina Dusevic who were assigned to this book had rendered their fullest support in putting the materials together and acted on all my requests in a timely manner.

Dr. Viduranga Y. Waisundara
Australian College of Business and Technology,
Kandy Campus, Kandy, Sri Lanka

Section 1

Fundamentals of Dietary Fibers

Chapter 1

Introductory Chapter: The Basics of Dietary Fibers

Kanchana Samarasinghe, Chamodya R. Dharmadasa and Viduranga Y. Waisundara

1. Introduction

The purpose of this chapter is to set the stage to the remainder of the content of the book by providing an overview about dietary fibers, their uses, benefits, chemistry and biochemical aspects. While the chapters are specific, in-depth and detailed, the contents of this particular chapter are kept general, so that readers of the book may gain an understanding of dietary fibers at the very beginning—especially if they are not familiar with this area of study, and go into a deeper understanding thereafter.

2. A brief overview of dietary fibers

Dietary fibers are considered a non-digestible form of carbohydrates, due to the inability of the digestive enzymes to break them down into monomeric units. The amount of dietary fiber in the diet varies based on the type of food and quantity consumed by an individual. The solubility of fibers has a significant impact on their function and therefore, they are classified into two main categories as shown in **Figure 1**.

Dietary fibers that are insoluble, such as cellulose, lignin and hemicellulose have distinct characteristics such as filaments resembling threads, and most insoluble dietary fibers such as numerous vegetables have a rough feel. Dietary fibers that are insoluble are less prone to fermentation as compared with soluble dietary fibers [1]. Sources of soluble and insoluble dietary fibers are shown in **Table 1**. The solubility of some of the commonly sourced dietary fibers is shown in **Table 2**. An overview of non-starch polysaccharides which come under dietary fibers is shown in **Table 3**.

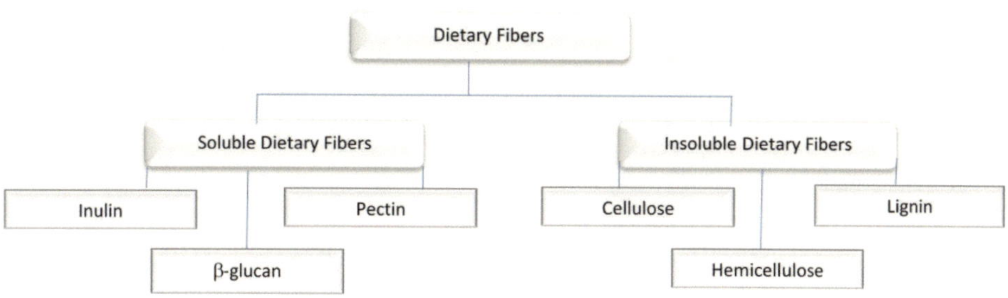

Figure 1.
The classification of dietary fibers (insoluble and soluble dietary fibers).

	Soluble dietary fibers			Insoluble dietary fibers	
Main constituent	Pectin	Gums	Lignin	Cellulose	Hemicellulose
Common food source	Citrus fruits	Oats	Strawberries	Green pepper	Bran cereals
	Apples	Cauliflowers	Pears	Peas	Mustard greens

Table 1.
Chemical constituents and common food sources of dietary fibers.

Polysaccharides / food Ingredients	Solubility
Locust beans—commonly extracted from carob trees	Only soluble in warm water
Carrageenan (Iota type)	Soluble both in room temperature and warm water
Carbohydrate gum also generally known as hydroxypropyl methylcellulose (HPMC)	Soluble in water in room temperature, but not in warm water

Table 2.
Common dietary fibers and their solubility.

Non-starch polysaccharides	Dietary fibers
Polyglucose	Beta-glucan, cellulose and hemi-cellulose
Polyfructose	Inulin

Table 3.
Overview of non-starch polysaccharides and their form of dietary fibers.

3. Dietary fibers of importance

Out of all soluble and insoluble fibers, there are some which are of high importance from a nutritive as well as an industrial perspective which are introduced in brief as below.

3.1 Inulin

Inulin acts as a storage polysaccharide. Inulin has been a component of the human diet since ancient times. According to BeMiller [2], ancient humans have consumed a variety of tubers and roots, some of which contain starch and others which contain inulin or a related gluco-fructan, on all five continents. Thus, it is very likely that incorporation of inulin in the modern-day diet is of importance as well due to consistency and exposure.

3.2 Pectin

The proportion of methylation, or the fraction of carboxyl groups bonded with methanol, is used to categorize pectin products. Pectins are utilized for a range of jellies and jams in the food industry, due to their unique ability to produce spreadable gels [2].

3.3 Beta-glucan

Beta-glucan has been identified as a supportive food ingredient in lowering cholesterol levels in the blood [3]. In addition, studies have showed that beta-glucan has the capability to reduce stress level at occasions [4, 5]. The variety of beta-glucan structures depends on environmental sources. It is a key component of plant cell walls. Cereal-glucan has been associated with lowering the risk of heart disease, whereas yeast-glucan has been observed to primarily boost the immune system's ability to fight cancer and infections [4, 5].

3.4 Cellulose

As the most well-known dietary fiber of all categories, there are several health benefits associated with cellulose including reduction of constipation, decrease in the incidence of diverticulitis, and weight reduction. Owing to its abundance in plant sources, cellulose is the most common form of dietary fiber that is present in the mammalian diet and, therefore, is also the most researched dietary fiber component of all.

3.5 Hemicellulose

Xylose, pentose sugars, and arabinose are components of indigestible hemicellulose [6]. These fibers absorb water while being insoluble and are found in a variety of fruits and vegetables.

3.6 Lignin

Lignin is thought to have gained its name from the Latin word "lignum," meaning "wood." Lignin is composed of phenolic compounds that are covalently linked to polysaccharides. As lignin-rich food are indigestible, they have a distinct feel than other food and are sometimes referred to as "woody." Lignin is an insoluble dietary fiber that helps prevent the formation of bile stones and lower cholesterol [7]. It also has aromatic properties.

4. Health benefits of dietary fibers: a summary

Dietary fibers are a very complicated collection of food components. The consumption of dietary fiber as a whole has several health advantages as it is linked to lower risk of diabetes, gastrointestinal problems, stroke, and hypertension [3, 4]. In the small intestine, dietary fibers have an inherent ability to bind to cholesterol and bile acids, and this ability is hypothesized to be the cause of the hypolipidemic action [5]. The high fiber content of vegetables, food rich in protein, various fruits and whole grains make them appealing targets for the prevention of diseases such as cardiovascular disease and atherosclerosis. Due to the general importance of food fibers, a huge and prospective market for fiber-rich foods and components has developed within in the recent years, and there has been a tendency in recent years to identify specific sources of dietary fiber that could be utilized in the food industry [8]. Soluble dietary fibers are currently used as an ingredient in beverages, and it is becoming a global trend in the food industry at present replacing traditional and other food products that are less in nutritional value [9].

In the study of colon cancer formation, the effects of fiber rich diets on microbial population and fecal matter sterol ratios in the colon have received much interest at present. This is mainly due to the vital properties of dietary fibers [10]. Among the shared effects of soluble and insoluble dietary fibers, weight loss is one of the main benefits that one can experience in addition to decreasing energy density and inflammation [11]. In addition, dietary fibers prevent syndromes such as bowel syndrome and metabolic syndrome [12]. Adequate dietary fiber is necessary for the human body to operate properly and to be healthy, free of NCDs and other diseases and conditions. It is critical that individuals pay closer attention to their daily fiber intake since fiber plays a vital role throughout the human body systems.

5. Physiochemical properties of dietary fiber

Dietary fiber is a polysaccharide mixture with several functional abilities that activates while it moves further in the mammalian gastrointestinal tract. These actions are a result of its physiochemical structure. Some structural properties of dietary fiber that helps in its optimal functionality are discussed in the subsequent sections.

5.1 Particle size and bulk volume

Digestion itself means breaking down of complex molecules into simpler structures. Equally, for dietary fiber, the size of the molecule plays a major role in exhibiting its primary functions such as the time taken for fermentation, bacterial degradation and hydration process.

An experiment conducted on the period of time used for coconut residue to fully undergo the hydration process showed that reduction of particle size from 1127 μm to 550 μm enabled them to retain the hydration properties [13]. In addition, it was observed that the capacity of fat absorption had increased.

5.2 Surface area

The geometrical linkage is an important factor of any molecule, especially in the instance of dietary fiber as it is capable of resisting digestion unlike other starch-based molecules. A major reason for this is its characteristic higher surface area.

The availability of dietary fiber to microbial breakdown in the colon is influenced by porosity and accessible surface, whereas the regiochemistry of the surface layer may play a role in some cases. Moreover, physiochemical characteristics (certain types of adsorption or binding molecules) are also responsible for some of the physiological effects of fiber in the diet. The porosity of the material and the surface area available to bacteria, use of molecular probes such as enzymes or other molecular probes are all dependent on its surface area. However, the fiber's architecture is an aspect that is linked to its origin and history of processing [14].

5.3 Solubility and viscosity

Solubility has a significant impact on the functionality of fiber. It is widely known that soluble, viscous polysaccharides can obstruct nutrient digestion and absorption in the stomach. The polymer is anticipated to be more energetically stable in the solid state than in solution if the polysaccharide structure allows molecules to fit together in a crystalline array [15].

5.4 Cationic binding and organic molecules

While charged polysaccharides (which including pectins *via* their carboxylic acid groups) and related compounds such as phytates in wheat fibers have been demonstrated to bind metal ions *in vitro*, fiber has been associated with limiting absorption of nutrients. Charged polysaccharides have little influence on mineral and trace element absorption, although related compounds such as phytates can be harmful. The capacity of different fibers to sequester and even chemically bind bile acids has been proposed as a possible mechanism by which dietary fibers rich in uronic acids and phenolic compounds may have a hypocholesterolemic effect [16].

5.5 The process of hydration

The hydration characteristics of dietary fiber influence their fate in the digestive tract and explain some of their physiological consequences. Swelling and water retention capacity give a broad picture of fiber hydration and are relevant for fiber-fortified diets. Bulk density reveals more about the fiber, including the substrate pore volume. It contributes to the existing knowledge of how fiber behaves in food and during the gastrointestinal transit. The physical processes that the fibers go through before entering the body alter the physical characteristics of the fiber matrix as well as the hydration properties [17].

6. Physiochemical mechanism of dietary fiber

Dietary fiber metabolizes only through the process of microbial fermentation. It is a plumber of carbohydrates. This is essentially due to the lack of enzymatic matter that is essential for breaking the glycosidic bond. The mammalian gastrointestinal (GI) tract consists of gut micro-biota, which is capable of fermenting. As the fiber moves through the GI tract, the gut micro-biota metabolizes it through fermentation.

Different species of gut bacteria act upon these large polymers of glucose in order to break them into monomers. Primary degraders such as *Bifidobacterium*, *Bacteroides* and *Ruminococcus bromii* that have the ability to ferment glucose from glucose polymers act upon these large polymers of glucose in order to break them into monomers [18].

Due to the lack of enzymes that initiate the cleavage of glucose from glucose polymers, a secondary degrader names *Firmicute* species rely on primary degraders to release glucose monomers. This process takes place in the lumen. After degradation the glucose short-chain fatty acid (SCFA—by-product of fermented dietary fiber), specifically butyrate is absorbed by the colon epithelial cells (bolonocytes). Later, the remaining SCFA enters the circulation by the portal vein (blood that transports to liver) [19].

7. Concluding remarks

The study of dietary fibers remains a vital area of study when it comes to food, nutrition and health. They have been added to a variety of food products as a whole as a food ingredient that imparts several health benefits, especially when it comes to the GI tract and the gut microbiome. As a highly studied area, it is hoped that the contents of this chapter has provided a general understanding of this food component and thereby demonstrated its essentiality as a nutrient, bioactive and ingredient.

Author details

Kanchana Samarasinghe, Chamodya R. Dharmadasa and Viduranga Y. Waisundara*
Australian College of Business and Technology – Kandy Campus, Kandy, Sri Lanka

*Address all correspondence to: viduranga@gmail.com

References

[1] Papathanasopoulos A, Camilleri M. Dietary fiber supplements: Effects in obesity and metabolic syndrome and relationship to gastrointestinal functions. Gastroenterology. 2010;**138**(1):65-72

[2] BeMiller JN. Carbohydrate chemistry for food scientists. Woodhead Publishing and AACC International Press, USA; 2018. https://doi.org/10.1016/C2016-0-01960-5

[3] Anderson JW, Baird P, Davis RH, Ferreri S, Knudtson M, Koraym A, et al. Health benefits of dietary fiber. Nutrition Reviews. 2009;**67**(4):188-205

[4] Ciudad-Mulero M, Fernández-Ruiz V, Matallana-González MC, Morales, P. Dietary fiber sources and human benefits: The case study of cereal and pseudocereals. In Advances in food and nutrition research. 2019;**90**:83-134. Academic Press.

[5] Cui J, Lian Y, Zhao C, Du H, Han Y, Gao W, et al. Dietary fibers from fruits and vegetables and their health benefits via modulation of gut microbiota. Comprehensive Reviews in Food Science and Food Safety. 2019;**18**(5):1514-1532

[6] Stanhope KL, Goran MI, Bosy-Westphal A, King JC, Schmidt LA, Schwarz J-M, et al. Pathways and mechanisms linking dietary components to cardiometabolic disease: Thinking beyond calories. Obesity Reviews. 2018;**19**(9):1205-1235. DOI: 10.1111/obr.12699

[7] Naumann S, Haller D, Eisner P, Schweiggert-Weisz U. Mechanisms of interactions between bile acids and plant compounds—a review. International Journal of Molecular Sciences. 2020;**21**(18):6495

[8] Dhingra D, Michael M, Rajput H, Patil RT. Dietary fibre in foods: A review. Journal of Food Science and Technology. 2012;**49**(3):255-266. DOI: 10.1007/s13197-011-0365-5

[9] Gunenc A, Hosseinian F, Oomah BD. Dietary fiber-enriched functional beverages in the market. In: Diet Fibre Funct Food Nutraceuticals From Plant to Gut. USA: John Wiley & Sons Ltd., 2016. pp. 45-75. https://doi.org/10.1002/9781119138105.ch3

[10] Sánchez-Alcoholado L, Ramos-Molina B, Otero A, Laborda-Illanes A, Ordóñez R, Medina JA, et al. The role of the gut microbiome in colorectal cancer development and therapy response. Cancers. 2020;**12**(6):1406

[11] Weickert MO, Pfeiffer AF. Impact of dietary fiber consumption on insulin resistance and the prevention of type 2 diabetes. The Journal of Nutrition. 2018;**148**(1):7-12. DOI: 10.1093/jn/nxx008

[12] Soliman GA. Dietary fiber, atherosclerosis, and cardiovascular disease. Nutrients. 2019;**11**(5):1155. DOI: 10.3390/nu11051155

[13] Raghavendra SN, Ramachandra Swamy SR, Rastogi NK, Raghavarao KSMS, Kumar S, Tharanathan RN. Grinding characteristics and hydration properties of coconut residue: A source of dietary fibre. Journal of Food Engineering. 2006;**72**:281-286

[14] Guillon F, Auffret A, Robertson JA, Thibault JF, Barry JL. Relationships between physical characteristics of sugar beet fibre and its fermentability by human fecal flora. Carbohydrate Polymers. 1998;**37**:185-197

[15] Guillon F, Champ M. Structural and physical properties of dietary fibres, and consequences of processing on human physiology. Food Research International. 2000;**33**:233-245

[16] Dongowski G, Ehwald R. Properties of dietary preparations of the cellan-type. In: Guillon F et al., editors. Proceeding of the PROFIBRE Symposium, Functional properties of non-digestible carbohydrates. Nantes: Imprimerie Parentheses; 1998. pp. 52-54

[17] Thibault JF, Lahaye M, Guillon F. Physiochemical properties of food plant cell walls. In: Schweizer TF, Edwards CA, editors. Dietary fibre, a component of food. Nutritional function in health and disease: Springer-verlag, Berlin; 1992. pp. 21-56

[18] Macfarlane GT, Englyst HN. Starch utilization by the human large intestinal microflora. Journal of Applied Bacteriology. 1986;**60**:195-201. DOI: 10.1111/j.1365-2672.1986.tb01073.x

[19] Scheppach W, Luehrs H, Menzel T. Beneficial health effects of low-digestible carbohydrate consumption. The British Journal of Nutrition. 2001;**85**(Suppl 1):S23-S30. DOI: 10.1079/bjn2000259

Chapter 2

Psyllium: A Source of Dietary Fiber

Ranu Agrawal

Abstract

Dietary fiber is commonly known as roughage. Fibers are mostly present in vegetables, whole grain, nuts, legumes, and fruits. This is an indigestible part of the food obtained by plants. It includes polysaccharides such as cellulose, hemicellulose, pectic substances, mucilages, gums and lignin as well. Dietary fiber has beneficial physiological effect on health, so it is included in daily diet to decrease occurrence of several diseases. In this sequence, this chapter describes about the dietary fiber, psyllium commonly known as Isabgol which is prepared from the seed of the *Plantago ovata Forsk* (Psyllium ispaghula). Psyllium is hydrophilic mucilloid, has the capacity to absorb water and increases in volume while absorbing water. Psyllium consists of mixed viscous polysaccharide in which about 35% soluble and 65% insoluble polysaccharides (cellulose, hemicellulose, and lignin) are present. This can be used as gelling, food thickener, emulsifying and stabilizing agents in some food products. Psyllium is a natural biopolymer which has high quantity of hemicelluloses consist of xylan backbone connected with arabinose, galacturonic acid and rhamnose units. Since last many years it is being used as therapeutic agent in several diseases like chronic constipation, inflammation of mucous membrane of GIT tract, duodenal ulcers, piles or diarrohoea etc. It may be source of renewable and biodegradable polymer.

Keywords: Psyllium, hemicellulose, dietary fiber, therapeutic, lignin

1. Introduction

The term Dietary fiber was coined by Hipsley in 1953 who explained it as a plant cell wall constituent which was indigestible [1]. Later, in 1982, Kay defined the dietary fiber as a plant food component present everywhere and consists of substance having diverse morphological and chemical structure and also cannot be affected by human alimentary tract enzymes [2]. Dietary fiber was also defined by American Association of Cereal Chemists as the edible plant's part or analogue carbohydrate that are not digested or absorbed in the human's small intestine and partially or completely fermented in the large intestine [3]. There are several types of dietary fiber available. They may be soluble or insoluble types, or natural or artificial. Among all dietary fibers, psyllium is one of the important dietary fiber. From last several years it has been focused by various researchers because it contains beneficial pharmaceutical properties.

2. Psyllium occurrence

Psyllium is scientifically known as *Plantago* (family Plantaginaceae), a plant native to tropical regions. Psyllium word is commonly used for greater than 200 species of the *Plantago* genus. It is also known as isabgol and ispaghula in common Indian language. Psyllium was indigenous plant of Persia and Isabgol word has come from the Persian word "band ghoul" means "horse flower" which expresses the shape of Psyllium seed [4–6]. This plant is generally 10 to 18 inches short stemmed annual herb which is known by different names in different regional language such as isabgol, aspaghol, ashwagolam, bazarqutuna, aspagol, blond psyllium. This plant is extensively grown in many parts of the world. Normally, psyllium is cultivated for its mucilage substance, which is a white fibrous substance having hydrophilic characteristics. *Plantago ovata* and *Plantago psyllium* species are generally cultivated commercially to manufacture mucilage. The psyllium husk (seed coat) is also commercially employed in food industry as in bakeries, ice creams and candies. India plays a major role to make psyllium available in world market. In India, Gujarat and Rajasthan collectively have an area about sixty one thousand hectares for its cultivation. Some popular global brand names of psyllium are konsyl, modane bulk powder, bonvit, meta-mucilage, perdiem fiber, siblan, psyllium husk, serutan, fybogel etc.

2.1 Psyllium products

2.1.1 Psyllium seeds

Psyllium seeds are dried ripen seeds obtained from *Plantago ovata* plant. It is obtained by cleaning seeds from all dust, wastes, stones, agri farm fibers, iron particles and other impurities. Psyllium seeds are light brown in color and have faint odor. These seeds are made up of 40% of an essential fatty acid, Linoleic acid. Along with that, it may have 18.8% protein content, 19% fiber content and 10–20% triglycerides. It may contain soluble and insoluble fiber. Seed contains mucilage polysaccharide fiber, which is soluble in nature. Psyllium seeds are graded on the basis of purity and quality of the seeds.

2.1.2 Psyllium husk

The psyllium husk is actually the outer coating of seed which is made of mucilage around the seed. This is the main part of the plant which has nutritional value and used to manufacture psyllium products. This consists of proteins, polysaccharides, glycosides, vitamin B1 and choline. It contains high fiber content which is mainly composed of hemicellulose. Hemicellulose is a complex polysaccharide found in grains, vegetables and fruits, and although it is indigestible, but it is partially digested in colon part and nourishes to intestinal flora. Psyllium husk are obtained by processing of removal of outer coating of seed. It contains about 70% soluble fiber content and 30% insoluble fiber content. Husk is white fibrous substance and used in various industries like pharmaceutical, cosmetics and food product. There are various grades of psyllium husk available according to user needs on the basis of purity and mesh size. Psyllium husk is used as raw matter to form psyllium mucilage. It is precipitated with alcohol in aqueous solution and then washed with acetone and dried. Psyllium husk powder is also processed from the husk by using grinder of various particle mesh size.

2.2 Psyllium as a functional food

Psyllium is being used in various commercial industries such as in food, pharmaceutical and other industries. Psyllium is often added to functional food material due to its various physiochemical properties such as meal replacements, breakfast cereals, biscuits, bread and in bakery products. It is also added to shakes, juices, yogurt, syrups, soups and even in ice creams to improve the fiber content of the food. It is also used as a thickening agent in drinks or frozen desserts. Psyllium dietary fiber contains supplementary nutrients and phytochemicals which make it to be taken as whole food and close the fiber gap. The various efforts are going on to improve physiochemical, biological, sensory and functional properties of psyllium to enhance its utilization in food and safety. Psyllium has an extreme water absorbing capacity so it is not dispersed in water or aqueous solutions.

Psyllium has been focused by many researchers giving many approaches to improve its functionality. Food and drug administration (FDA) in 1998 recommended that cholesterol can be lower or attained in adequate amount by including lesser amount of fiber in daily diet. Mucilage present in seed husk has significant property as a thickener that can be used in food industries. It can be used as an ingredient of chocolate and other foods. It can be utilized as a main stabilizer in ice-cream. It improves the taste of mouth and feel of drinks or flavored drinks by making them more consistent and richer. Fiber supplements are demanded by consumers due to its pleasing taste and good storage stability. It can be employed as a foundation of cosmetics and for sizing functions. The psyllium husk has good binding capacity so it can be used as good binder and can be disintegrated in compressed tablets. Psyllium produces jelly like structure when it is treated with hot caustic soda which may be a substitute of agar-agar gel. Psyllium seed gum has been employed to prepare germicidal lubricating gel and dry dentifrice powder. It can also be used as a constituent of petroleum. It can be used as cattle feed and by mixing with jaggery in case of lactating cattle. The de-husked seed may be used as bird feed which is about 69% of the total seed crop weight. Husk has been considered very safe for use in nutraceutical and functional food. The use of psyllium husk in various food products has been approved by FDA [7]. Psyllium seeds have been used for many years in traditional medical prescription. Psyllium supplements are used as fiber formulation in high fiber consumer products including powder, granule, wafer, and capsule forms. Because psyllium has pharmacological effects, it is used to make fortified food.

Psyllium consists of a great amount of soluble fiber which fulfills daily dietary fiber recommendation of the body. The soluble fiber of psyllium effects the body lipids and proteins and related to metabolic processes. Bakery products are manufactured formulating different dosage of psyllium husk up to the possible level without causing harmful change in quality. The varying dosage of psyllium can also be given in hypercholesterolemia considering hormonal status in men and women as pre and post menopausal women. In postmenopausal women, about 15 gm daily dose can significantly decrease total cholesterol level and reduces the risk of coronary heart diseases whereas no significant decrease of total cholesterol was observed in premenopausal women. Psyllium dosage did not affect concentration of triglycerides, apolipoprotein A1 and apolipoprotein B in pre and postmenopausal women. In diabetic patients, higher glucose level and in hyperlipidemia due to more polysaccharide contents psyllium is added in processed food to control weight by its gel forming nature. The presence of bioactive substance arabinoxylan in fiber improves the quality of baking products and marked therapeutic potential. In the patient of celiac diseases, psyllium is added to bread dough with 93% acceptance rate whereas in non celiac disease patient acceptance rate is 97%. While making bread dough products having chemical composition less than 42.3% fat and less than 32% calories, psyllium can replace gluten in composition. By this way, products made with psyllium fiber contain less fat and lower calories. Psyllium fiber significantly increases water absorption determined by Farinograph (rheological device), simultaneously as increasing its amount. Psyllium contents increases the Falling number (FN) index of wheat flour by increasing its water absorption capacity, in addition to decrease in α amylase activity. Psyllium is studied for the development of "spongy dessert" by adding its mucilage powder which is extracted from the psyllium seeds. This mucilage powder is also added to prepare milk solid dessert made from low fat cow milk. The developed herbal dessert product may contain nutrients low in carbohydrate, high protein and dietary fiber, saturated fats, free from trans fats and lesser calories. It also has the properties to make relief from acidity and constipation.

Psyllium gums and mucilage are naturally occurring biopolymers. Their applications in pharmaceutical, neutriceutical and biotechnological fields are increasing day by day. Psyllium has been successfully used as a thickening agent, colloidal stabilizer, and as a gelling agent for past many years in the food industries as well as pharmaceutical industries. Psyllium mucilage characteristics make them unique for using as a matrix for delivery and/or for entrapment of different drugs types, proteins and cells. Being a natural polysaccharide it is very important in industrial applications.

Psyllium functions as prebiotics which is a substance required for the healthy microbial colonies of probiotics growing in the gut. These are gut microbiome which make an essential ecosystem of microorganism inside the colon. Healthy colony of microorganisms in gut is also necessary for strong immune system and makes your body efficient to fight against infection, maintain healthy cells and tissues and reduce inflammation. In addition to, prebiotics make easier bowel movement in patients suffering constipation. Psyllium husk support the gut microbiome and it is beneficial for gut microflora particularly in constipated patients [8]. Psyllium plays an important role to increase the production of short chain fatty acids such as propionate and butyrate that are required for microbial health. Another characteristic of psyllium is to retain water in the small intestine, in this way water flow increases into the ascending colon.

3. Characterization of Psyllium

Various researchers investigated morphology and thermal behavior of psyllium using spectrophotometric method like FT-IR and thermo-gravimetric analysis. The FT-IR spectrum of psyllium shows an absorption peak at 3401 cm^{-1} which attributes –OH (alcohol) stretching. An absorption band at 2926 cm^{-1} is credited to C-C stretching band of alkanes, on the other hand band at 1050 cm^{-1} is due to C-O-C stretching of ether representing polysacchrarides. The absorption band at 896, 714 and 613 cm^{-1} may be attributed to polymer chain bending [9].

Scanning electron microscopy study of psyllium reveals its surface structural morphology. The powder form of psyllium shows irregular shape matrix of unequal size which are structural constituents of fiber. There are not significant differences in physical structure of fiber and show irregular matrix of protein and fiber. At higher magnifications, the psyllium's complex ultra-structure shows a hard surface deficient with granular structures, which presents proteinaceous and fiber material of psyllium. Micrograph shows presence of small cavities or pores which may affect the physicochemical properties like water holding and oil absorption capacity of psyllium [9–11].

Thermal characterization reveals broad information about thermal transition and thermal stabilities behavior of psyllium. DSC (Differential scanning Calorimetry) technique is useful for providing thermodynamic property of conformation transition state of polysaccharides. During conformation transition in DSC measurements obtained endothermic peaks show melting of structural domains [11].

4. Physicochemical properties

The physiological properties of dietary fiber are associated mainly with its solubility, viscosity, water-holding capacity, bulking ability, binding ability, fermentability and so on [11, 12].

4.1 Solubility

According to solubility in water, dietary fiber may be of two type i.e. soluble fiber and insoluble fiber. This nature of dietary fiber makes them technologically and physiologically functional. Presence of soluble fiber content increases solubility of dietary fiber and reduce plasma cholesterol and glycemic response [13]. Insoluble fiber content provides porosity and low density which increase fecal bulk when ingested in diet and decrease in intestinal transit. In food processing procedures, use of soluble fiber in food products formulate them more beneficial because psyllium gives viscosity, which makes them able to form gel or they can also act as emulsifiers in comparison to insoluble fiber.

4.2 Viscosity and gel formation

Viscosity is a physicochemical characteristic which is associated with soluble dietary fiber contents like pectins, gums, and glucans. Viscosity and gel formation capacity is linked to soluble fiber's capacity of absorbing water and formation of gelatinous mass [14]. Soluble fiber forms gel and increases the viscosity of gastrointestinal tract contents. This phenomenon may clear the delayed gastric clearance often linked with fiber ingestion. This viscous nature of fiber also gives lubrication of stool [15].

4.3 Water-holding capacity

Water holding capacity (WHC) term is defined as the quantity of water which is held by known mass of dry fibers or hydrocolloid under certain conditions as temperature and time duration of soaking. In general, the polysaccharide contents of dietary fiber are hydrophilic in nature and water is retained on the hydrophilic sites on the fiber on surface or in void spaces of the fiber molecules [16].

4.4 Binding ability

Dietary fiber can trap bile acids secreted in small intestine by gall bladder. These soluble fibers make gel matrix that finally exit in feces. The physical entrapment comes into view in the terminal part of small intestine where bile acids are usually reabsorbed [17].

4.5 Fermentability

The fermentation ability of fiber is highly variable which ranges from non fermentable lignin to almost complete fermentable pectin. The fermentation of soluble fiber takes place to greater extent by colonic bacteria but insoluble fiber are not fermented. The ability of soluble fiber to be fermented makes the psyllium physiologically effective. Plants have different proportions of fiber on the basis of fermentability such as rapidly fermented, slowly fermented and unfermented fiber. Some fruits as apples and bananas and vegetables like beans and potatoes contain rapidly fermented fiber and may contribute less to bulk feces in comparison to other fiber. Psyllium and wheat bran are considered to ferment at slow rate at the entire length of colon and contribute more fecal mass.

4.6 Bulking ability

Insoluble fiber, like lignin and cellulose, generally may remain unfermented by microflora present in colon and take part to increase fecal bulk by forming particles and holding water. Wheat bran is considered as the best bulking agent. Some fermentable fiber as hemicellulose present in cabbage can increase fecal bulk by rising fecal flora. On the other hand, extremely fermentable fiber like pectin shows very little effect to increase fecal bulk.

5. Chemical composition

Psyllium fiber is viscous in nature and beneficial for human health, in prevention as well as treatment of diseases. Psyllium contains soluble and insoluble fiber contents. Psyllium fiber contains mainly mucilage, which is found in seed coat. The mucilage is extracted by grinding of outer coat of the seed. It is clear, colorless and gel forming agent. Psyllium is a mixture of polysaccharides, for example, hexoses, pentoses, and uronic acid. Mucilage is composed of about 15% non-polysaccharides matter such as fat and protein and remaining 85% yields a single polysaccharide containing D-xylose (~62%), L-rhamnose (~9%), L-arabinose (~20%), and D-galactouronic acid (~9%). The ß-D-xylose residues in the pyranose ring form a linear backbone of polysaccharide. Disaccharide side chain is linked with terminal α-D-galactouronic acid and O-2- of α-L-rhamnose. In the polymer backbone, all the three side chains are linked to either O-2 or O-3 of xylose. The backbone has both β (1>3) and β (1>4) glucosidic linkage (**Figure 1**). Total protein fraction is about

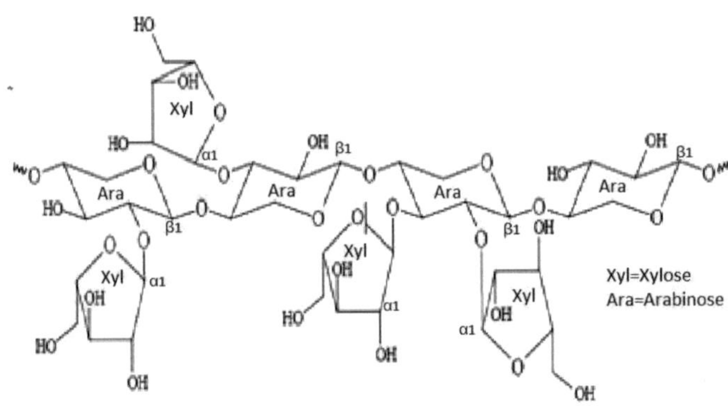

Figure 1.
Arabinoxylan structure.

globulin 23.9%, albumin 35.8%, and prolamin 11.7%. Psyllium seeds also yield oil which has high content of unsaturated fatty acids such as oleic acid (39.1%) and linoleic acid (40.6%) and small amount of linolenic acid (6.9%).

Psyllium husk also has a high quantity of hemicellulose which consists of a xylan backbone associated with rhamnose, arabinose, and galacturonic acid units (arabinoxylans). The seed is composed of 35% soluble fraction and 65% insoluble polysaccharides such as cellulose, hemicellulose, and lignin. Psyllium is a hydrophilic mucilloid and highly branched arabinoxylan polysaccharide consisting of high water holding and gel formation ability. Gel-forming fraction of the polysaccharides consists of xylose, arabinose, and other sugars in trace amount.

6. Drug delivery studies

Drugs are not often administered in the body as only pure chemical substances, but are generally given through drug delivery system (DDS). The DDS should consist of active pharmaceutical component in association with inert or excipients substances. Drugs are transformed to dosage forms by adding one or more substances which may be referred as excipients, essentially these materials must be pharmacologically inert. These ingredients may be used to achieve certain goals such as to modify appearance, improve handling, enhancing physical property like absorption, efficiency, retention time. In different kind of diseases such as chronic constipation, inflammation of mucous membrane of gastrointestinal and genito-urinary tracts, diarrhea, and duodenal ulcer, piles, gonorrhea, etc. Psyllium is used as bulk forming, demulcent, non-irritant laxative drug and as a cervical dilator etc. So, the psyllium plays an important role in drug delivery system or pharmaceutical application as well as dietary supplements.

A large number of polysaccharides containing excipients that are obtained from natural sources possess their own importance. They have different properties which makes them to be used in variety of applications like binding agents, suspending agents, coating materials, granulating agents, rising viscosity of aqueous solution and easily dispersible material in pharmaceutical industry. Natural occurring polysaccharides are hydrocolloid polymer, which can be used as gel forming substance, binder, sweetener, flavoring agent, taste masking agent, lubricants to make easy to swallow component. A recent need of this area is to study such natural occurring useful substances which tend to have the property of biodegradability, biocompatibility and non-toxicity.

Psyllium seeds and husk are broadly used in pharmaceutical application as an emollient laxative, demulcent, drug therapy to regulate lipid and glucose levels and various diseases. There are several drug delivery systems have been explored to deliver drug for the therapeutic purposes, for example hydrogel, which is made of natural polymer and proved as an excellent carrier for the drug molecule by controlling its release and target. Psyllium has been attracted by researchers due to having a neutral pH, extended transportation time and reduced enzymatic activity.

Psyllium acts as an anti-ulcer agent itself. Singh and coworker investigated that using psyllium with rabeprazole drug as drug carrier, may increase the drug therapeutic potential [18]. Psyllium fiber is one of the important gel having glucose dropping effect. Psyllium reduces hyperglycemia by inhibiting glucose absorption in intestine and also improves motility. As per pharmacological view, its polysaccharides have significance to decrease glucose absorption. Drug delivery strategies which are associated with hydrogel psyllium and properly modified for the synthesis of hydrogel can efficiently function as high potential candidate to get better drug delivery systems [19]. Anderson and colleaguesinvestigated use of psyllium for long term action of mild to moderate hypercholesterolemia [16]. Daily tolerate dose of psyllium was observed i.e., 11.5 gm and side effects were for short duration or minor or not related to treatment.

A grafted natural polymer of psyllium has been synthesized and used to formulate in different types of drug delivery system. The haemocompatibility of psyllium was checked by learning the blood relations with graft copolymer with context to thrombogenicity and haemolytic potential also. Thrombogenicity results indicated non-thrombogenic effect of graft copolymer as the thrombus percentage and weight of clot shaped of polymer was in lesser amount in comparison of positive control [20].

In recent studies, colon specific drug delivery systems have been significant. It is investigated that treatment of colon associated diseases requires more colonic concentration of drug in the colon. So, there should be a flexible approach for effective therapy to deliver drug to colon. In conventional therapy system does not complete pharmacokinetic profile, especially for the poisonous drugs having good therapeutic index. Therefore, in ideal situation a good profile can be achieved by using natural polysaccharides matrix which can preserve good therapeutic index as well as controlled drug delivery. The optimum pH for drug release medium was studied for drug loaded hydrogel at different pH medium. The amount of drug tetracycline hydrochloride, tyrosin and insulin were released maximum at pH 7.4 in comparison to pH 2.2 buffer and distilled water. The observation was considerable for colon specific drug delivery systems. Polysaccharides based drug delivery devices can be considerably better candidates than different hydrogels for controlling drug release [21]. It may be used as fabricated acrylic based graft copolymer for colon precise drug delivery. The model drugs as tetracycline hydrochloride, tyrosine and insulin dynamics were let out from tailored psyllium with acrylic acid cross linked copolymer drug laden hydrogel. The tailored psyllium was formed with methacrylamide poly (MAAm), 2-hydroxylethylmethacrylate (2-HEMA), acrylamide (AAm), poly(vinyl alcohol) and poly(acrylic acid) based polymeric networks. In this case, the released contents of water soluble drug get trapped in hydrogel when water moved in the network of swelled polysaccharides and drug get dissolved followed by aqueous pathways along with diffusion to the device surface. The amount of released drug was found more at pH 7.4 buffer solution.

Psyllium was studied for delivery of anticancer drug 5-fluorouracil [18]. Drug could be released in different pH buffer with psyllium and that was pH unresponsive. This drug was released from drug laden hydrogel (per gram) psyllium. The other drug tetracycline hydrochloride and rifampicin could be released from the tailored psyllium which was developed by radiation crosslinked polymerization with methacrylamide. Rifmapicin is a broad spectrum antibiotic used for the treatment of *Mycobacterium* contagions as tuberculosis and laprosy and also it is used against methicillin-resistant *Staphylococcus aureus* bacteria in combination with fusidic acid. Hydrogel composed of crosslinked macromolecular chain which makes entangled mesh structure, forming a matrix for the entrapment of the drugs. When drug laden matrix comes in touch with solvent, polymeric backbone of hydrogel become relaxed and liquefied drug get diffused into the outside discharging medium. The more swelling of gel makes the fast release of drug from the polymer matrix.

Thrombogenicity as well as haemolytic potential of psyllium hydrogels was studied for accessing the blood connection with graft copolymer and found there was very low percentage of clot and thrombus formed. The clot formation was lesser in the membrane in comparison of positive control so these polymer were classified as non-thrombogenic. Psyllium PVA-hydrogel may be partial hemolytic as studied suggested. Psyllium-poly(acrylic acid) was studied for electrical stimulus sensitive drug delivery system. The swelling property of psyllium hydrogel polymer was measured comparing with artificial biological fluid. Maximum swelling was observed in artificial biological fluid. On electrolysis, rapid swelling was observed due to dissociation of solvent system into ions which made easy entrance into the gel network. Swelling property of graft copolymer was a main function of chemical architecture.

7. Pharmaceutical uses

Psyllium is naturally found swellable biomaterial. It has been used as a traditional medicine since a long time. It is currently being of great interest for utilizing in pharmaceutical industry. Seeds of psyllium are used to thicken tablets and capsules at the time of manufacturing, because it contains hexoses, pentoses and uronic acids. Some studies reported antioxidant properties of psyllium which may be very effective for the treatment of inflammatory bowel disease. Phytochemical studies revealed that biological activity of psyllium attributes to the presence of some secondary metabolic products such as aucubin glycosides, flavanoids, fixed oil contents, tannin, sugar, sterols and proteins as well as hydrocolloidal mucilage which is present in the outer seed coat [22]. It has been found most effective to cure various physiological disorders for example, it reduces blood cholesterol levels thus decrease risk factors of cardiovascular system diseases. Cardiovascular diseases are the number one cause of death globally. Earlier studies showed that psyllium seeds affect the immune system of rabbits by reducing anti-HD antibodies and increases white blood cells amount in the blood as well as spleen leucocytes. It has been also shown the extract of psyllium may be used in diabetes treatment. It can control blood sugar level by reducing hyperglycemic condition in type 1 and type 2 diabetes. This dietary fiber is also used as fiber supplements to regulate bowel function in intestine.

8. Therapeutic benefits

Psyllium is much more beneficial for healthy life. It is extensively considered safe, medically secure and efficient for treatment of certain type of diseases when recommended dose is taken.

8.1 In hepatic diseases

In cell, due to over generation of free radical oxidative stress developed which may lead to cell damage resultantly various disorders can happen as liver dysfunction. The natural antioxidants present in psyllium as polyphenols and flavanoids may prevent cells from oxidative stress and damage. So, it may act as free radical scavenger and may be used to cure various disorders. Mekky and coworkers investigated antioxidant activity of psyllium seeds to protect against CCl_4 (free radical) induced-hepatotoxicity experimentally [23]. It also has enhanced endogenous antioxidant capability of hepatic tissue and inhibited lipid peroxidation.

8.2 Bowel diseases

Psyllium fiber supplementation may be useful to cure irritable bowel diseases, inflammatory bowel disease and ulcerative colitis. This useful effect is most likely associated with its anti-constipation action. Psyllium fiber when reaches to intestine, is digested by anaerobic fermentation and resultantly short chain fatty acids like butyrate, acetate and propionate are produced, which have antioxidant and anti inflammatory property. Increased concentration of short chain fatty acids may yield high energy to colonic mucosa because they act as a substrate for oxidation. So along with constipation activity increased level of short chain fatty acids have helpful effects on ulcerative colitis and inflammatory bowel disease. Chaplin and colleague showed psyllium supplementation which might be applicable in diseases like hemorrhoids and diverticulitis [24]. It may be beneficial to cure hemorrhoids by reducing bleeding when come in contact and of congested hemorrhoidal cushions [25]. Psyllium dietary fiber supplement regulates bowel function. It may reduce the risk of diseases like diabetes, obesity, and certain gastrointestinal disorders by taking sufficient quantity in daily diet.

8.3 Gastrointestinal disorders

Most dietary fiber sources activate laxation which increase conic contents and stimulate propulsion in intestine. Fiber undergoes for anaerobic fermentation and incompletely fermented or unfermented fibers associate with moisture holds and increases mass of stool. These fibers also function as substrate for microbial growth and this additionally bacterial mass also increases total colonic content [16]. Unfermented gel of psyllium fiber also functions as an emollient and lubricant which leads to easy passage of the stool movement. Psyllium fiber is used widely as a fiber supplement to cure constipation as it increases moisture level and total mass of dry stool. It has been proposed that 1 gram of psyllium fiber can increase 5.9–6.1 gram of stool weight. This was more effective in comparison with oat bran fiber and wheat bran fiber. Psyllium fiber has the great ability to hold water so it has also been shown to slow the time of gastric empty and colon transit. This is the opposite of the preferred effect against constipation but it is beneficial for persons with diarrhea or uncontrolled fecal defecation with liquid stools [26].

8.4 Cardiovascular diseases

Psyllium fiber intake in daily diet reduces risk of coronary heart diseases. It was approved by the US Food and Drug Administration (USFDA) in 1998 under the Nutrition Labelling and Education Act which was linking with study of psyllium fiber. Psyllium fiber has the properties to lower cholesterol. LDL-cholesterol and total cholesterol are recognized as biomarkers or risk factors for heart disease. Program in food safety, nutritional and regulatory affairs (PFSNRA) in 2006 suggested that 7 grams of psyllium soluble fiber intake in diet resulted significant physiologically reduction of LDL-cholesterol which ranges from 0.047% to 0.86% per gram fiber basis.

9. Conclusion

Psyllium is obtained from the seed of the *Plantago ovata Forsk* (Psyllium ispaghula), a rich source of dietary fiber. Psyllium consists of mixed viscous polysaccharide in which about 35% soluble and 65% insoluble polysaccharides (cellulose, hemicellulose, and lignin) are present. Psyllium is a natural biopolymer which has high quantity of hemicelluloses. It may be a source of renewable and biodegradable polymer. The physiological properties as solubility, viscosity, water-holding capacity, bulking ability, binding ability, fermentability make psyllium effective for the use as medicine and functional food. Psyllium is much more beneficial for healthy life. It is extensively considered safe and secure medically and efficient for treatment of certain type of diseases such as hepatic diseases, gastrointestinal diseases and cardiovascular diseases. Psyllium has been currently being of great interest for utilizing in pharmaceutical industry.

Conflict of interest

The author declares no conflict of interest.

Author details

Ranu Agrawal
Sir Chhotu Ram Institute of Engineering and Technology (SCRIET), Chaudhary Charan Singh University, Meerut, India

*Address all correspondence to: ranu6_78@yahoo.com

References

[1] Hipsley E. H. Dietary fibre and pregnancy toxaemia. Br Med J. 1953;22(2):420-422. DOI:10.1136/bmj.2.4833.420

[2] Kay R. M. Dietary fiber. J Lipid Res. 1982;23:221-242.

[3] AACC. Report of the Dietary Fiber Definition Committee to the Board of Directors of the American Association of Cereal Chemists 2001;46(3):112-126.

[4] Ricklefs-Johnson K, Johnston CS, Sweazea KL. Ground flaxseed increased nitric oxide levels in adults with type 2 diabetes: A randomized comparative effectiveness study of supplemental flaxseed and psyllium fibre. Obesity Medicine. 2017;5:16-24. DOI:10.1016/j.obmed.2017.01.002

[5] Sukhija S, Singh S, Riar CS. Analyzing the effect of whey protein concentrate and psyllium husk on various characteristics of biodegradable film from lotus (*Nelumbo nucifera*) rhizome starch. Food Hydrocoll. 2016;60:128-137. DOI:10.1016/j.foodhyd.2016.03.023

[6] Yu L, Yakubov GE, Zeng W, Xing X, Stenson J, et al. Multi-layer mucilage of Plantago ovata seeds: Rheological differences arise from variations in arabinoxylan side chains. Carbohydr Polym. 2017;165:132-141. DOI:10.1016/j.carbpol.2017.02.038

[7] Seeds ARE. Dietary fiber: Role in nutrition management of disease. In: Guide to Nutritional Supplements. Caballero, B. (ed.). Academic Press, USA (2009).

[8] Jalanka J., Major G., Murray K., Singh G., Nowak A., Kurtz C., Santiago I. S., Johnston J. M., de Vos W.M., and Spiller R. The effect of psyllium husk on intestinal microbiota in constipated patients and healthy controls. Int J Mol Sci. 2019;20(2):433-444. DOI:10.3390/ijms20020433

[9] Kaith BS, Kumar K. Vacuum synthesis of psyllium and acrylic acid based hydrogels for selective water absorption from different oil–water emulsions. Desalination 2008;229:331-341. DOI:10.1016/j.desal.2007.08.020

[10] Sen G, Mishra S, Rani GU, Rani P, Prasad R. Microwave initiated synthesis of polyacrylamide grafted Psyllium and its application as a flocculant. Int J Biol Macromolec 2012;50:369-375. DOI:10.1016/j.ijbiomac.2011.12.014

[11] Shah AR, Sharma P, Gour VS, Kothari SL, Dar KB, Ganie SA, and Shah YR. Antioxidant, nutritional, structural, thermal and physico-chemical properties of psyllium (*Plantago ovata*) seeds. Current Research in Nutrition and Food Science 2020;8(3):727-743. DOI:10.12944/CRNFSJ.8.3.06

[12] Kumar D, Pandey J, Kumar P, Raj V. Psyllium mucilage and its use in pharmaceutical field: An overview. Curr Synthetic Sys Biol 2017;5:134-140. DOI:10.4172/2332-0737.1000134

[13] Brennan CS. Dietary fibre, glycaemic response, and diabetes. Mol Nutr Food Res 2005;49(6):560-570. DOI:10.1002/mnfr.200500025

[14] Sierra M, García JJ, Ferna'ndez N, Diez MJ, Calle AP, Farmafibra group. Therapeutic effects of psyllium in type 2 diabetic patients. Eur J Clin Nutr 2002;56:830-842. DOI:10.1038/sj.ejcn.1601398

[15] Marteau P, Flourie B, Cherbut C, et al. Digestibility and bulking effect of ispaghula husks in healthy humans. Gut 1994;35:1747-1752. DOI:10.1136/gut.35.12.1747

[16] Anderson JW, Allgood LD, Turner J, et al. Effects of psyllium on glucose and serum lipid responses in men with type 2 diabetes and hypercholesterolemia. Am. J. Clin. Nutr. 1999;70:466-473. DOI:10.1093/ajcn/70.4.466

[17] Purohit P, Rathore HS. Isabgol: Aherbal remedy. World J Pharm Res. 2019;8(7):579-585. DOI:10.20959/wjpr20197-15001

[18] Singh B, Chauhan N, Kumar S, Bala R. Psyllium and copolymers of 2-hydroxylethylmethacrylate and acrylamide-based novel devices for the use in colon specific antibiotic drug delivery. Int J Pharm. 2008;352:74-80. DOI:10.1016/j.ijpharm.2007.10.019

[19] Thakur VK, Thakur MK Recent trends in hydrogels based on psyllium polysaccharide: A review. J Clean Prod. 2014;82:1-5. DOI:10.1016/j.jclepro.2014.06.066

[20] Kaith BS, Kumar K. Preparation of psyllium mucilage and acrylic acid based hydrogels and their application in selective absorption of water from different oil/water emulsions. Iran Polym J. 2007;16:529

[21] Singh B. Psyllium as therapeutic and drug delivery agent. Int J Pharm. 2007;334:1-4. DOI:10.1016/j.ijpharm.2007.01.028.

[22] Tewari D, Nishat A, Tripathi Y. (2014). World J Pharm Res 3:361-372. DOI:0.13140/2.1.5043.4885

[23] Mekky M. A., Shaymaa M. M., Ahmed W., Amr E. A., Ahmed M. O., Hanan A. S., Thagfan S A L, Atta E.Z. A study of the hepatoprotective effect of *Plantago psyllium* L. seed extract against carbon tetrachloride induced hepatic injury in rats. J Appl Biomed. 2020;18/2-3:80-86. DOI:10.32725/jab.2020.006

[24] Chaplin MF, Chaudhury S, Dettmer PW, Sykes J, et al. Effect of ispaghula husk on the faecal output of bile acids in healthy volunteers. J Steroidal Biochem. Mol. Biol. 2000;72:283-292. DOI:10.1016/s0960-0760(00)00035-2

[25] Kritchevsky D. Bile acids: Biosynthesis and functions. Eur. J. Cancer Preven. 1991;1(2):23-28.

[26] Owen RW, Henly PJ, Day DW, Thomson MH, et al. Fecal steroids and colorectal cancer: Bile acid profiles in low and high risk groups. In: Human Carcinogenesis. Joossens JV, Hill MJ, Geboers J. (Eds.). Elsevier, Amsterdam, 1985; 165-170. PMID:1842728

Section 2

Dietary Fibers and Disease Prevention

Chapter 3

Dietary Fiber and Dyslipidemia

I. Gusti Ayu Nyoman Danuyanti and Z.S. Ahmad Fahrurrozi

Abstract

Fibers are abundantly found in vegetables, fruit, beans, cereals, seeds, and tubers. Beans and seeds, alongside prevailing as both of the fiber sources, are the sources of vegetable protein as well. Whereas tubers are a carbohydrate source, which people deem as a staple food. Fiber intake in diets, particularly soluble fibers, has the ability to produce gel in the intestines, inhibiting glucose and cholesterol absorption. Dietary fibers have the ability to bind bile salts in the digestive tract, and disturbed bile reabsorption will stimulate bile synthesis in the liver. Dyslipidemia has a significant role in systemic responses and inflammation in adipose tissues. Inflammation can increase intestinal permeability and adipose tissues. Dyslipidemic management is carried out by altering lifestyles, intervening in suitable diets to reduce LDL levels, and increasing HDL levels. The degree of compliance with diet interventions is seminal to ensure successful dyslipidemic management.

Keywords: Fiber, Dyslipidemia, Management

1. Introduction

Dyslipidemia as one of the risk factors of metabolic syndrome is the abnormality condition of lipid profile, marked by the increased triglyceride levels (TG), total cholesterol, low density lipoprotein (LDL) cholesterol, and low level of high density lipoprotein (HDL) cholesterol. Dyslipidemia is triggered by lifestyle changes with the tendency of consuming high fat but low fiber food and sweet beverages with high level of fructose alongside with the lack of physical activity [1–4].

High fat dietary with saturated fatty acid and trans fatty acid substance initiates the rising of LDL cholesterol, reduces the level of DHL cholesterol, and arouses oxidative stress on endothelium blood vessel as the result of over production of reactive oxygen species (ROS) that will oxidize extracellular LDL, developing oxidized LDL [5, 6]. On the other hand, high fructose dietary initiates insulin resistance through the reduction of insulin receptor sensitivity [7, 8]. Insulin functions as expression control of sterol regulatory element binding protein (SREBP), roles in the regulation and biosynthetic of fatty acid and cholesterol in liver [1, 9, 10]. As a result, it can increase SREBP expression and appropriately stimulates liver lipogenesis and triglyceride synthesis enhancement in liver.

2. Dietary fiber and dyslipidemia

2.1 Fiber

Fibers are mostly found in food with a low glycemic index [11], so the higher the fiber level contained in the food, the lower the glycemic index. This is because fibers bring the food bolus into a more viscous condition (gel-formed), thereby slowing down the food digestion process [12–14].

Fibers are abundantly found in vegetables, fruit, beans, cereals, seeds, and tubers. Beans and seeds, alongside prevailing as two fiber sources, are also the sources of vegetable protein, whereas tubers are a carbohydrate source, which people deem as a staple food. An example of fiber sources from tubers is sweet potatoes (*Ipomoea batatas*), which contain fibers of 3–4.2% [15, 16]. Meanwhile, fibers in sweet potato starch are 5.54% [17]. In addition to sweet oranges, yellow pumpkins belong to the vegetable group possessing beta carotenes and considerable high fibers. Previous research attested that yellow pumpkin starch contained fibers by 10–12.24%, while fresh yellow pumpkins contained fibers by 2–3% [15, 18, 19].

Fibers, by definition, are carbohydrate polymers which indigestibled in small intestines but are managed to be fermented by bacteria in the colon [11]. Following their characteristics, fibers are classified into eight, as listed in detail in **Table 1**.

Referring to the Dietary Reference Intake (DRI), fibers are classified into three [12, 21]:

1. Dietary fibers are carbohydrates and lignin, which in terms of intrinsic, intact in plants, and indigestible.

2. Functional fibers are carbohydrates, which indisgestibled but beneficial for humans.

3. Total fibers are a combination of dietary fibers and functional fibers.

Food products with 2.5 g of fibers/portion are considered as a good source of fibers, and those with 5 g of fibers/portion are considered an excellent one [20, 22]. Dietary fibers are not correlate with energy building, however, after experiencing fermentation in the colon, the fibers were capable of increasing the volume of feces, enhancing laxative products, softening stool consistency, and forming a short-chain fatty acid (SCFA) contributive to health [11, 23, 24].

2.1.1 Dietary fiber

Dietary fibers are a part of plants, which consumabled and collated from carbohydrates and lignin in plants, resistant to the digestive and absorption in the small intestines and experiencing a partial or simultaneous fermentation in the colon (Brownlee, 2009). The sources of dietary fibers are not only vegetables and fruit but also beans, cereals, seeds, and tubers [24, 25].

The fermentation process, undergone by dietary fibers in the colon, broke down dietary fibers into SCFA, giving the physiological functions, beneficial for health [26]. Primary short-chain fatty acids created by acetate, butyrate, and propionate simultaneously contributing to the mineral absorption process, fat metabolism, and anti-inflammation [26]. The degree of fiber fermentation which creates SCFA products is presented in **Table 2**.

Some factors which affected fiber fermentation were types of substrate (the chemical structure of fibers and solubility), specific microbes (gut microbial activities and population), and transit time in the digestive tract [27, 28].

Butyric acid is needed in maintaining the balance of colon cells by increasing the growth and differentiation of cells and demonstrates higher anti-inflammation than acetate and propionate. Acetate is imperative in increasing ileal motility as well as blood flow to the colon and increasing lipopolysaccharides in relation to Tumor Necoris Factor (TNF), Interleukin-6 (IL-6), and Nuclear

Classification	Types of fiber
1. Dietary fibers	Lignin
	Cellulose
	β-glucan
	Hemicellulose
	Pectin
	Gums
	Resistant starch
2. Functional fibers	Resistant dextrin
	Psyllium
	Fructooligosaccharides
	Polydextrose
	Isolated gums
	Isolated resistant starch
3. Soluble fibers	B-glucan
	Gums
	Wheat dextrin
	Psyllium
	Pectin
	Inulin
4. Insoluble fibers	Cellulose
	Lignin
	Several types of pectin
	Several types of hemicellulose
5. Fermentable fibers	
6. Non-fermentable fibers	Wheat dextrin
	Pectin
	β-glucan
	Guar gum
	Inulin
	Cellulose
	Lignin
7. Viscous fibers	Pectin
	β-glucan
	Guar gum
	Psyllium
8. Non-viscous fiber	Polydextrose
	Inulin

Table 1.
Classification by-characteristic of Fibers [20].

Sources of fibers	Fermentation (%)
Mono/oligosaccharides	
Pectin	
Bran	
Cellulosa	
Lignin	

Table 2.
Degree of Fiber fermentation [27, 28].

Factor Kappa Beta (NFκβ) by increasing periphery antibody production. Propionic fatty acid decreased food intake and increased satiety by suppressing leptin activities and activating G protein-coupled receptor-43 and -41 (GPCR-43 and GPCR-41) [22, 24, 26].

Dietary fibers, classified by the solubility properties, were soluble dietary fibers and insoluble dietary fibers [20, 25, 26]. Soluble dietary fibers contributed to forming thick solution (viscous), slowing down gastric emptying and absorption of nutrients including glucose so as to control plasma glucose levels, whereas insoluble dietary fibers functioned to overcome digestive tract disorders, increase the volume of feces, and shorten the transit time of feces in the colon [14, 20].

2.1.2 Soluble dietary fiber

Inulin, resistant starch, and soluble polysaccharides are soluble dietary fibers. Inulin is a fructooligosaccharide (FOS) class of carbohydrates important to the digestive system, increase the number of intestinal L cells in the proximal part of the colon, elevate SCFA production, decline the pH of the colon so inhibit the growth of pathogenic bacteria, increase the volume of feces, and avert constipation. Inulin is customarily used in food industries as a substitute for fat and sugar, specifically in low-fat food products [14, 29, 30].

Resistant starch is kind of starch or a starch-degraded product undigested in the human small intestine [31] and could be categorized into four [32]:

1. Resistant starch 1 (RS1) is starch whose resistance formed after encapsulated in an indigestible matrix.

2. Resistant starch 2 (RS2) is starch whose granules are not gelatinized so they were slowly hydrolyzed by α-amylase.

3. Resistant starch 3 (RS3) is starch formed when food containing a cooked and cooled starch.

4. Resistant starch 4 (RS4) is starch which has been chemically modified.

Starch is composed of amylose and amylopectin which could experience gelatinization when being cooked and is easily hydrolyzed by amylase enzymes [22, 31, 33].

Water-soluble polysaccharides (WSP) were water-soluble dietary fibers as a plant's component which could not be enzymatically degraded into a subunit which could be absorbed in the stomach and small intestine. Water-soluble polysaccharides were primarily used by food industries to achieve food quality in regard to viscosity, stability, texture, and appearance [34].

Dioscorea contained glucomannan, including in the WSP group which could develop and thicken in water [14, 34].

2.1.3 Insoluble dietary Fiber

Insoluble dietary fibers comprised carbohydrates which contained cellulose, hemicellulose, and non-carbohydrates with lignin [20, 33]. Insoluble dietary fibers were more frequently found in vegetables, wheat, cereals, and beans. They have several functions, e.g., increasing the volume of feces and shortening the transit time of feces in the colon. Because of these functions, they were often exerted to treat digestive tract disorders, such as constipation, diverticular diseases, and irritable bowel syndrome [24, 25, 33].

What is contained by insoluble dietary fibers tremendously determines the physiological effects of the fibers. For instance, cellulose and hemicellulose retained water in feces, enhanced peristalsis of the colon, escalated the colon performances, and reduced the colon intraluminal pressure; whereas lignin is physiologically pivotal in binding minerals, increasing the secretion of bile acids, and acting as an antioxidant [24, 35].

2.1.4 Functional fibers

Functional fibers were also insoluble dietary fibers which were indigestible and has beneficial effects as some other dietary fibers has different psychological effects [22, 25]. It is because rich-fiber food also contained bioactive phytochemicals which has additional benefits [33, 36]. Components of dietary fibers considerably determined psychological effects bred. Psychological functions of dietary fibers could bring on the occurrence of some diseases protective effects, e.g. [32]:

1. Giving glycemic controls and insulin responses to patients with diabetes mellitus.

2. Improving intestinal health through the functions of fibers as prebiotics and protagonist cultures in patients with colon cancer, ulcerative colitis, inflammatory bowel disease (IBD), diverticulitis, and constipation.

3. Fixing blood lipid profile in cardiovascular disease and dyslipidemia.

4. Increasing satiety and reducing energy intake in obesity.

5. Increasing micronutrient absorption in patients with osteoporosis.

2.2 Fiber and dyslipidemia

One of the indicators of dyslipidemia is an increase in LDL-cholesterol levels and a decrease in LDL-cholesterol levels so the first target of a dyslipidemic therapy is LDL management. Dyslipidemic management is carried out by altering lifestyles, intervening in suitable diets to reduce LDL levels, and increasing HDL levels [14, 37]. The degree of compliance with diet interventions is seminal to ensure successful dyslipidemic management. Recommended dietary patterns for dyslipidemia are listed in **Table 3** [14, 37].

No.	Nutrition	Recommended intake
1.	Total fat	25–35% of the total calorie
2.	Saturated fat	Less than 7% of the total calorie
3.	Trans fatty acid	0
4.	Polyunsaturated acid	Maximal 10% of the total calorie
5.	Monounsaturated acid	Maximal 20% of the total calorie
6.	Carbohydrate	50–60% of the total calorie
7.	Fiber	25–30 g/day
8.	Protein	±15% of the total calorie
9.	Cholesterol	Less than 200 mg/day
10.	Total calorie (energy)	Energy intake and expenditure should be in balance to maintain the desired body weight and prevent weight gain.

Table 3.
Recommended dietary patterns for patients with Dyslipidemia [14, 37].

The recommended dietary patterns should be in control using the following indicators [14, 24, 28]:

1. An increase in saturated fat in the diet by 1% would increase LDL by about 2%. A decrease in saturated fat intake would decrease LDL levels by 8%.

2. High saturated fat intake correlated with high coronary heart disease within a population.

3. Trans fatty acid is able to increase serum LDL levels so it would be better not to consume or consume it at a minimum intake level.

4. If monounsaturated, fatty acid changed saturated one, it is able to reduce LDL but not HDL or increase triglycerides.

5. If polyunsaturated fatty acid changed saturated one, it is managed to reduce LDL and, at a high dose, HDL too.

6. If saturated fat and trans fat intake are minimal, restricting total fat intake is unnecessary.

7. Carbohydrate intake should be less than 60% to reduce risks of decreased HDL and increased serum triglycerides.

8. Consuming 5–10 g/day of soluble fibers is managed to decrease LDL levels by 5%.

Fiber intake in diets, particularly soluble fibers, has the ability to produce gel in the intestines, inhibiting glucose and cholesterol absorption [38]. Dietary fibers have the ability to bind bile salts in the digestive tract, and disturbed bile reabsorption will stimulate bile synthesis in the liver. Cholesterol is a precursor to bile synthesis, so increased bile synthesis would decrease cholesterol levels in the blood [39]. Short-chain Fatty Acid (SCFA), yielded from fiber fermentation in the colon, is also substantive in inhibiting the activity of hydroxymethyl GLUTaryl-CoA reductase (HMG-CoA reductase) enzyme which is needed in forming mevalonates as the primary product for cholesterol synthesis [24, 38, 40].

Dyslipidemia has a significant role in systemic responses and inflammation in adipose tissues. Inflammation can increase intestinal permeability and adipose tissues. A dyslipidemic condition also influences the composition of intestinal microflora and has a direct impact on body weight. Intestinal microbes were the primary sources of several molecules, e.g., lipopolysaccharides and peptidoglycan, as the causes of inflammation in periphery tissues [41]. An increase in Short-chain Fatty Acid (SCFA) in the intestines could reduce the intestinal pH, which indirectly altered the composition of intestinal microflora by cutting the number of harmful/pathogenic bacteria. In terms of the inflammatory process brought about by dyslipidemia, butyrates and propionates suppressed NFκβ \activation through the inhibitory pathway for the activation of the IκB Kinase (IKK) protein complex, decreasing synthesis and cytokine secretion and proinflammatory adhesion molecules [28, 42, 43].

Previous studies argued that propionates could reduce free-plasm fatty acid levels through a lipolysis inhibition process in adipose tissues, reducing TLR4 receptor expressions to be able to bind with a free fatty acid, which consequently reduced TLR4 expression which could stimulate NFκβ activation and translocation. The ratio of butyrate effects to inhibition in an inflammatory process is higher than propionates [43, 44].

2.3 Fiber, dyslipidemia and its relationship with autoimun disease

Inflammation is one of the immune response which if it does excessive can cause autoimmune diseases. Inflammation is often followed by an increase in ESR, CRP, cytokines and complement activation which is able to result in tissue damage. Dyslipidemia is a potential trigger for chronic inflammation. Thereby, dyslipidemia is associated with autoimmune diseases. Although it is not clear whether dyslipidemia is a predisposing factor or an outcome of autoimmune disease. In line with this, an obesity is managed to increase the incidence and severity of several autoimmune diseases, such as psoriasis and rheumatoid arthritis. Cytokines in the form of IL-17a are known to be pathogenic in psoriasis and RA. For this reason, IL-17a blocking is used as the treatment of autoimmune diseases [45].

Furthermore, the autoimmune condition plays a role in the development of dyslipidemia and atherosclerotic plaque formation in patient with autoimmune rheumatoid disease (ARD), such as RA. This is related to the formation of autoantibodies and chronic inflammation which occured. As in SLE, ox-LDL is also frequently find in synovial biopsy specimens of RA patients. The product of ox-LDL is able to be recognized by the scavenger receptor and influence the action of macrophages. Meanwhile, the results of ox-LDL digestion by macrophages can be toxicable to endothelial cells, chemotaxis of inflammatory cells and cause changes in smooth muscle function. In the state of dyslipidemia, especially hyperlipidemia, serum LDL levels are elevated [46].

Additionaly, increasing endogenous butyrate production is managed to be a valuable strategy in the prevention of obesity and related metabolic diseases. However, in the other side, this also can increase exogenous intake through butyrate supplements. Most likely, the causative lack of randomized controlled trials proving the efficacy of butyrate in these metabolic disorders is mainly due to the poor palatability of the actual butyrate preparations available on the market. Nevertheless, there is an urgent need for products that mask the unpleasant organoleptic properties of butyrate, in oder to facilitate clinical studies in children and in adult patients [47].

Increasing interest in the effect of dietary fiber, on lowering the blood lipid concentration. There are various mechanisms by which serum and hepatic lipids are reduced by dietary fiber: binding to bile, viscosity, and bucking in the small intestine caused the suppression of glucose and lipid absorption, increased production of SCFAs, and modulation of lipid metabolism-related genes. In addition, dietary fibers, classified as the seventh nutrients, are generally considered safe, but over-consumption could cause intestinal discomfort. From the above evidences, dietary fibers could be used as alternative supplements to exert health benefits, including lipid-lowering effects on humans. However, more clinical evidence is needed to strengthen this proposal and its fully underlying mechanism still requires more investigation. Only if we fully understand the mechanism and dose relationship of each kind of DFs we are able to apply them in the intervention of hyperlipidemic patients [48].

In populations who habitually consume diets rich in plant foods, great adherence to three types of plant-based diets were differentially associated with risk of incident dyslipidemia. Study result strongly supports considering the quality of plant foods for dyslipidemia prevention. Prospective studies are needed to confirm the relationship between a plant-based diet and dyslipidemia in diverse populations with different dietary habits [49].

3. Conclusions

Dyslipidemia has an important role in systemic and inflammatory responses in adipose tissue. Inflammation can increase the permeability of the intestines and adipose tissue. One indicator of dyslipidemia is an increase in LDL cholesterol levels and a decrease in LDL cholesterol levels so that the first target of dyslipidemia therapy is LDL management. Dyslipidemic management is carried out by changing the lifestyle, intervening in an appropriate diet to reduce LDL levels, and increasing HDL levels. The level of adherence to dietary interventions is critical to ensure successful dyslipidemic management. The recommended dietary pattern should be controlled using the following indicators. Increasing saturated fat in food by 1% will increase LDL by about 2%. Decreasing saturated fat intake will reduce LDL levels by 8%.

Acknowledgements

This chapter is supported by Politeknik Kesehatan Mataram and Badan Pengembangan dan Pemberdayaan Sumberdaya Manusia, Ministry of Health Republic of Indonesia.

Notes/thanks/other declarations

Thank you for using this reference as input for the management of dyslipidemia.

Author details

I. Gusti Ayu Nyoman Danuyanti* and Z.S. Ahmad Fahrurrozi
Department of Medical Laboratory Technology, Politeknik Kesehatan Kemenkes Mataram, Mataram, Indonesia

*Address all correspondence to: danuyanti@gmail.com

References

[1] Avramoglu, R.K., Basciano, H. & Adeli, K. Lipid and lipoprotein dysregulation in insulin resistant states. Clin Chim Acta. 2006, 368(1-2): 1-19.

[2] Coate, K.C., Scott, M., Farmer, B., Moore, M.C., Smith, M., Roop, J., Neal, D.W., Williams, P. & Cherrington, A.D. Chronic consumption of a high-fat/ high-fructose diet renders the liver incapable of net hepatic glucose uptake. Am J Physiol Endocrinol Metab. 2010, 299(6): E887–E898.

[3] Fernández-Sánchez, A., Madrigal-Santillán, E., Bautista, M., Esquivel-Soto, J., Morales-González, Á., Esquivel-Chirino, C., Durante-Montiel, I., Sánchez-Rivera, G., Valadez-Vega, C. & Morales-González, J.A. Inflammation, oxidative stress, and obesity. Int J Mol Sci. 2011, 12(5): 3117– 3132.

[4] Zhu, Y., Wang, C., Song, G., Zang, S., Liu, Y. & Li, L. Toll-like receptor-2 and-4 are associated with hyperlipidemia. Mol Med Rep. 2015, 12(6): 8241– 8246.

[5] Tuminah, S. Efek Asam Lemak Jenuh dan Asam Lemak Tak Jenuh' Trans' Terhadap Kesehatan. Media Penelit dan Pengembang Kesehat. 2009, XIX(II): S13–S20.

[6] Welinsa, F., Asni, E. & Malik, Z. Histopatologi Aorta Torasika Rattus Novergicus Strain Wistar Jantan Setelah 8 Minggu Pemberian Diet Aterogenik. JOM. 2015, 2(1): 1-11.

[7] Prahastuti, S. Konsumsi Fruktosa Berlebihan dapat Berdampak Buruk bagi Kesehatan Manusia. JKM. 2011, 10(2): 170-174.

[8] Tranchida, F., Tchiakpe, L., Rakotoniaina, Z., Deyris, V., Ravion, O. & Hiol, A. Long-term high fructose and saturated fat diet affects plasma fatty acid profile in rats. J Zhejiang Univ Sci B. 2012, 13(4): 307-317.

[9] Basciano, H., Federico, L. & Adeli, K. Fructose, insulin resistance, and metabolic dyslipidemia. Nutr metab. 2005, 2(1): 1-14.

[10] Jameel, F., Phang, M., Wood, L.G. & Garg, M.L. Acute effects of feeding fructose , glucose and sucrose on blood lipid levels and systemic inflammation. Lipids Health Dis. 2014, 13(1): 1-7.

[11] Jung, C.-H. & Choi, K.M. Impact of High-Carbohydrate Diet on Metabolic Parameters in Patients with Type 2 Diabetes. Nutrients. 2017, 9(4): 322.

[12] Jenkins, D.J.A., Marchie, A., Augustin, L.S.A., Ros, E. & Kendall, C.W.C. Viscous dietary fibre and metabolic effects. Clin Nutr Suppl. 2004, 1(2): 39-49.

[13] Jung, U.J. & Choi, M.-S. Obesity and its metabolic complications: the role of adipokines and the relationship between obesity, inflammation, insulin resistance, dyslipidemia and nonalcoholic fatty liver disease. Int J Mol Sci. 2014, 15(4): 6184-6223.

[14] Sunarti. Serat Pangan. Yogyakarta: Gadjah Mada University Press. 2017

[15] Mahmud, M.K., Hermana, Zulfianto, N.A., Apriyantono, R.R., Ngadiarti, I., Hartarti, B., Bernadus & Tinexcelly. Tabel Komposisi Pangan Indonesia. PT Elex Media Komputindo, Jakarta. 2009

[16] Helen, O.T., Olusola, A.E., E, I.E., Bond, A.U. & Ezekiel, A.D. Ipomoea batatas L . Extract Reduces Food Intake , Fasting Blood Glucose Levels and Body Weight. Europ J Med Plants. 2013, 3(4): 530-539.

[17] Ambarsari, I., Sarjana & Choliq, A. Rekomendasi Dalam Penetapan Standar

Mutu Tepung Ubi Jalar. Standardisasi. 2009, 11(3): 212-219.

[18] Guine, R.P.., Pinho, S. & Barroca, M. Study of the convective drying of pumpkin (*Cucurbita maxima*). Food and Bioprod Process. 2011, 89(2011): 422-428.

[19] Noelia, J., Roberto, M.M., Jesús, Z.J. De & Alberto, G.J. Physicochemical , technological properties , and health-bene fi ts of *Cucurbita moschata* Duchense vs . Cehualca A Review. FRIN. 2011, 44(9): 2587-2593.

[20] Slavin, J. Fiber and prebiotics: mechanisms and health benefits. Nutrients. 2013, 5(4): 1417-1435.

[21] Kendall, C.W.C., Esfahani, A. & Jenkins, D.J.A. The link between dietary fibre and human health. Food Hydrocoll. 2010, 24(1): 42-48.

[22] Clark, M.J. & Slavin, J.L. The effect of fiber on satiety and food intake: A systematic review. J Am Coll Nutr. 2013, 32(3): 200-211.

[23] Kusharto, C.M. Serat makanan dan perannya bagi kesehatan. Jurnal Gizi dan Pangan. 2006, 1(2): 45-54.

[24] Brownlee, I.A. The physiological roles of dietary fibre. Food Hydrocolloids. 2009, 25(2011): 238-250.

[25] Santoso, I.A. Serat pangan (dietary fiber) dan manfaatnya bagi kesehatan. Magistra. 2011, 23(75): 35-40.

[26] Zeng, H., Lazarova, D.L. & Bordonaro, M. 2014. Mechanisms linking dietary fiber, gut microbiota and colon cancer prevention. World J Gastrointest Oncol. 2014, 6(2): 41-51.

[27] Cook, S.I. & Sellin, J.H. Short chain fatty acids in health and disease. Aliment Pharmacol Ther. 1998, 12(6): 499-507.

[28] Tan, J., McKenzie, C., Potamitis, M., Thorburn, A.N., Mackay, C.R. & Macia, L. The Role of Short-Chain Fatty Acids in Health and Disease. Adv Immunol. 2014, 121: 91-119.

[29] Kolida, S., Tuohy, K. & Gibson, G.R. Prebiotic effects of inulin and oligofructose. Br J Nutr. 2002, 87(S2): S193–S197.

[30] Pompei, A., Cordisco, L., Raimondi, S., Amaretti, A., Pagnoni, U., Matteuzzi, D. & Rossi, M. In effectvitro comparison of the prebiotic effects of two inulin-type fructans. Anaerobe. 2008, 14(5): 280-286.

[31] Leszczyñski, W. Resistant starch– classification, structure, production. Poli. J. Food Nutr Sci. 2004, 13(54): 37-50.

[32] Fuentes-Zaragoza, E., Riquelme-Navarrete, M.J., Sánchez-Zapata, E. & Pérez- Álvarez, J.A. Resistant starch as functional ingredient: A review. Food Res Int. 2010, 43(4): 931-942.

[33] Lattimer, J.M. & Haub, M.D. Effects of Dietary Fiber and Its Components on Metabolic Health. Nutrients. 2010, 2: 1266-1289.

[34] Prabowo, A.Y., Estiasih, T. & Purwantiningrum, I. Umbi Gembili (*Dioscorea esculenta* L.) Sebagai Bahan Pangan Mengandung senyawa Bioaktif. JPA. 2014, 2(3): 129-135.

[35] Kumar, V., Sinha, A.K., Makkar, H.P.S., De Boeck, G. & Becker, K. Dietary roles of non-starch polysachharides in human nutrition: A review. Crit Rev Food ScI. 2012, 52(10): 899-935.

[36] Lairon, D. Dietary fiber and control of body weight. NMCD. 2007, 17: 1-5.

[37] Mahan, L.K., Escott-Stump, S. & Raymond, J.L. Krause's Food & the Nutrition Care Process, 2012.

[38] Maryanto, S., Fatimah, S., Sugiri, S. & Marsono, Y. Efek Pemberian Buah Jambu Biji Merah terhadap Produksi Scfa dan Kolesterol dalam Caecum Tikus Hiperkolesterolemia. Agritech. 2013, 33(3): 334-339.

[39] Gérard, P. Metabolism of cholesterol and bile acids by the gut microbiota. Pathogens. 2014, 3(1): 14-24.

[40] Post, R.E., Mainous, A.G., King, D.E. & Simpson, K.N. Dietary fiber for the treatment of type 2 diabetes mellitus: a meta-analysis. JABFM. 2012, 25(1): 16-23.

[41] Tremaroli, V. & Backhed, F. Microbiota and host metabolism. Nature. 2012, 489: 242-249.

[42] Tedelind, S., Westberg, F., Kjerrulf, M. & Vidal, A. Anti-inflammatory properties of the short-chain fatty acids acetate and propionate: a study with relevance to inflammatory bowel disease. World J Gastroenter. 2007, 13(20): 2826.

[43] Al-Lahham, Peppelenbosch, M.P., Roelofsen, H., Vonk, R.J. & Venema, K. Biological effects of propionic acid in humans; metabolism, potential applications and underlying mechanisms. Biochim Biophys Acta. 2010, 1801(11): 1175-1183.

[44] Dahl, W.J., Agro, N.C., Eliasson, A.M., Mialki, K.L. & Olivera, J.D. Health Benefits of Fiber Fermentation Health Benefits of Fiber Fermentation. J Am Coll Nutr. 2017, 36(2): 127-136.

[45] L. R. Augustemak de Lima et al., "Dyslipidemia, chronic inflammation, and subclinical atherosclerosis in children and adolescents infected with HIV: The PositHIVe Health Study," PLoS One, vol. 13, no. 1, pp. e0190785–e0190785, Jan. 2018, doi: 10.1371/journal.pone.0190785.

[46] E. Matsuura, F. Atzeni, P. Sarzi-Puttini, M. Turiel, L. R. Lopez, and M. T. Nurmohamed, "Is atherosclerosis an autoimmune disease?," BMC Med., vol. 12, no. 1, pp. 1-5, 2014.

[47] S. Coppola, C. Avagliano, A. Calignano, and R. Berni Canani, "The Protective Role of Butyrate against Obesity and Obesity-Related Diseases," Molecules, vol. 26, no. 3, p. 682, Jan. 2021, doi: 10.3390/molecules26030682.

[48] Y. Nie and F. Luo, "Dietary Fiber: An Opportunity for a Global Control of Hyperlipidemia," Oxid. Med. Cell. Longev., vol. 2021, 2021.

[49] K. Lee, H. Kim, C. M. Rebholz, and J. Kim, "Association between Different Types of Plant-Based Diets and Risk of Dyslipidemia: A Pros

Chapter 4

Suitability of Fruits and Vegetables for Provision of Daily Requirement of Dietary Fiber Targets

Oluwafunmilayo Dorcas Adegbaju, Gloria Aderonke Otunola and Anthony Jide Afolayan

Abstract

The risk factors associated with low dietary fiber intake and the synergy with its role in colon prebiotic activity has stimulated a re-awakening in the scientific research. Dietary fiber intake has reduced all over the world, and so it has been labelled as a major shortfall nutrient of important in public health. Changes in lifestyle and improved standard of living have affected the diet of consumers in so many ways. Observation of these facts have spurred a special interest in the search for functional foods that contains essential nutrients like dietary fiber whose nutritional value improves the health of the consumer, enhances their physical and mental state and prevent lifestyle diseases. Fruits and vegetables are a modest source of total dietary fiber with nutrients such as vitamins, minerals, and phytochemicals, including polyphenols, which provide support for their biological plausibility and enhance their health benefits. This chapter therefore reviews existing literature on the utilization of fruits and vegetables as rich sources of fiber; their fiber concentration, their appropriateness in meeting the adequate fiber intake for daily consumption and their overlapping roles as a fiber source and as nutraceuticals.

Keywords: Dietary fiber, chronic diseases, fruits, vegetables, functional fiber, adequate intake

1. Introduction

Diets of most industrialized nations have become sophisticated, transiting from the nutritious traditional diet to high-energy density and low nutrient diversity diet. This is largely reinforced by changes in lifestyle and improved standard of living, which in turn has resulted into an upsurge in lifestyle diseases such as diabetes [1], obesity, hypertension [2], intestinal cancer [3], obstructive sleep apnea and cardiovascular disease, caused by imbalanced diet [4]. In addition to the premature-mortality cases recorded for these diseases, they are also known to have a great impact on psycho-social functioning [5], work productivity [6], and global healthcare expenditure [7, 8]. These concerns have stimulated a special interest in the search for functional foods which benefits the consumer's physical and mental state; and with the ability to prevent lifestyle diseases. Research in this regard has occurred primarily in western societies where a high prevalence of these chronic diseases has been observed and ultimately address the more frequent and severe malnutrition among lower-income countries [9].

Significant advances have been made recent years regarding the search for functional foods for metabolic regulations and medical therapeutic approaches to various chronic diseases. However, regardless of therapeutic choice for the management of most of these diseases, ultimately, the solution stems from behavioral change at an individual level [4]. One of the important factors that can guarantee an overall healthy lifestyle of an individual, is the consumption of adequate, but not excessive, levels of nutrient. While daily nutritional requirements for every individual varies depending on sex, size, age, and activity levels [10, 11]. An individual's exact requirement for a specific nutrient is generally unknown, with multiple confounding factors such as variations in genetic, metabolic, and gut microbial factors. All these factors combine to create much uncertainty regarding the optimal dietary needs for the individual [4]. Therefore, assessment of dietary adequacy for an individual can be a cumbersome task, due to the ambiguities associated with estimating an individual's usual intake and the lack of knowledge of an individual's actual nutrient requirements [11].

In this chapter, current knowledge on:

1. Dietary fiber definition

2. Dietary fiber intake among western population

3. Recommended adequate intake for total fiber.

4. Dietary fiber food sources: fruits and vegetables for daily fiber target provision.

5. Health promoting properties of dietary fiber.

2. Dietary fiber

Since the mid-twentieth century, when the term "dietary fiber" was coined by Hipsley [12], there has been concern about accurate and meaningful definition of this macronutrient component of the human diet. Due to the complexity of its varying composition in different food, it was formerly referred to as the non-digestible component of plant cell wall that resist digestion by secretions of the human alimentary tract (cellulose, hemicelluloses, pectin, and lignin) [12]. In 1998, 1998 American Association of Cereal Chemists International (AACCI) referred to Dietary fiber as the edible parts of plants or analogous carbohydrates that are resistant to digestion and absorption in the human small intestine with complete or partial fermentation in the large intestine. Like the AACCI definition, Codex Committee on Nutrition and Foods for Special Dietary defined dietary fiber as carbohydrate polymers with a degree of polymerization not lower than 3 which are neither digested nor absorbed in the small intestine [13]. Its definition was further stretched to include indigestible plant material that are not cell-wall components such as gums, algal polysaccharides, mucilage, and carrageenan were also included as dietary fiber [14]. Analytically, dietary fiber is a non-starch polysaccharide with three or more monomeric units and lignin from plants. Lignin, being a complex polymer of phenylpropane residues; the remaining dietary fiber components are polysaccharides. These polysaccharides resist digestion because they are non-α-linked-glucan-polysaccharides, whereas the human digestive tract appears to secrete only α-glucosidases [15]. The European Food Safety Authority (EFSA) defined dietary fiber as non-digestible carbohydrate plus lignin. An extensive list of substances that constitute dietary fiber, such as hydrocolloids, non-starch polysaccharides, cellulose, fructo-oligosaccharides (FOS), and other resistant oligosaccharides were provided by EFSA [16].

Furthermore, dietary fiber is known to be the combination of "dietary" and "functional" fiber. It is classified into two basic categories based on water solubility: soluble fiber (pectins, gums and mucilage) and insoluble fiber (cellulose, hemicellulose, and lignin). This is because their physiological benefit differs based on their sources. However, most dietary sources of naturally occurring fiber contain both soluble and insoluble fiber, but in varying amounts [17]. Soluble fibers are known to be an active compound in the regulation of digestion and absorption in the gut and are known to be present in fruits, vegetables, legumes, sugar beet, potato, seed plants and seaweed extracts. In contrast, insoluble fibers are less fermented and will primarily act in the large intestine where it effectively increases fecal weight and volume, dilutes colonic contents, and decreases intestinal transit time. Its main food sources are vegetables, sugar beet, various bran, cereal grains, and wood plants [18–20]. Functional fiber refers to fiber sources (nondigestible carbohydrates) that are either synthesized, extracted, or isolated and manufactured from natural sources and been reported to show health benefits. They include β-glucans, cellulose, chitins, and chitosan, fructans, gums, lignin, pectin, polydextrose and polyols, psyllium, resistant dextrin, and resistant starches [21, 22]. Consumption of total fiber is considered as the sum of dietary fiber and functional fiber.

2.1 Dietary fiber intake among western population

The beneficial health effects associated with dietary fiber consumption and its synergy with the role of human intestinal microbiota has caused a re-awakening in the scientific research of its prebiotic properties [17]. Food composition tables from many countries now contain values for the dietary fiber content of foods. Several observational studies have established the link between fiber intake and the risk reduction of coronary heart disease, stroke, hypertension, diabetes, and diverticular diseases [23–25]. Low fiber intake is perhaps the most studied dietary risk factor responsible for the development of most cardiovascular diseases and other diverticular diseases [26].

In 2014, the Food Survey Research Group Dietary (FSRG), recorded the mean dietary fiber intake of all individuals two years and older, in US population, to be 15 grams per day (excluding breastfed children). While intakes of males and females were reported to be18 and 15 grams per day, respectively. The 2015–2020 Dietary Guidelines for Americans named fiber as a major shortfall nutrient of important public health concern [27, 28]. Despite the efforts in the past years to promote sufficient fiber consumption through fruit, vegetable, whole-grain, and other fiber sources, fiber intake has remained low when compared to the adequate recommended amount. Blacks had significantly lower dietary fiber intake (13 gm) compared to Whites (16 gm) and Hispanics (17gm) [29]. The National Health and Nutrition Examination Survey of 2009–2010 (NHANES), recorded vegetables and fruits as the highest contributors to the dietary fiber intake of the US population. They both accounted for the over one-quarter (28%) of the population intake (**Table 1**) [29].

The World Health Report of 2016 acknowledged high intake of saturated fatty acids, high total fat intake and inadequate consumption of dietary fiber as being among the world's most serious health risk factors [30]. In Australia, Dietary

Food categories	Individuals reporting (%)‡	Contribution to dietary fiber (%)
Vegetables: fresh, frozen, and canned vegetables, including potatoes	67	16
Fruits: fresh, frozen and canned fruit	48	12
Breads, rolls, tortillas: yeast breads and rolls, flour and corn tortillas, bagels, English muffins	66	12
Grain-based mixed dishes: pasta dishes, macaroni and cheese, burritos, tacos, tamales, fried rice,	33	9
Cereals: cooked and ready to eat cereals	33	8
Plant-based protein foods: beans, peas, legumes, nuts and seeds, processed soy products	28	8
Sweet bakery products, candy, other desserts: cakes, pies, cookies, snack/meal bars, ice cream	65	7
Savory snacks and crackers: potato chips, tortilla and corn chips, popcorn	46	6
Pizza: fast food/restaurant and frozen pizza	14	4

Data are from National Health and Nutrition Examination Survey (NHANES) 2009–2010 "What We Eat in America" (Fiber intake of the U.S population).
‡Percentage of individuals reporting the foods in the category at least once on the reporting day.
†Food categories not listed including soups, milk and dairy, burgers, and meat, poultry, seafood mixed dishes, contributed 3% or less to fiber intake.
†SOURCE: Hoy and Goldman [29].

Table 1.
The 2011–2012 National Nutrition and physical activity survey, showing vegetables as the food category with the highest contribution to dietary fiber consumption.

fiber is regarded as a nutrient of concern in diets, especially among adolescents, young adults and generally among the lower socio-economic status individuals falling short of recommendations. From the 2011–2012 National Nutrition and Physical Activity Survey, only 28.2% of children met the adequate fiber intake and less than 20% of adults met the suggested dietary target to reduce the risk of chronic disease [31]. The survey also showed that fruits and vegetables contributed greatly to dietary fiber intake (**Table 1**), as they accounted for 28% of population intake.

2.2 Recommendation for adequate intake (AI) level for dietary fiber

The compelling evidence of the health benefits associated with the consumption of dietary fiber has led to the emphasis on the increment of its inclusion in daily diet. According to the Institute of Medicine "the recommended Dietary Reference Intake (DRI) daily allowance for individuals aged 50 years and younger is 25 to 38 g/day (14g/1,000 kcal/day). For men aged 19–50 years the daily recommendation is 38 g/day and women 25 g/days, and for men ages > 51 is 31 g/day and women ages > 51 is 21 g/day. The recommendation for children ages 1–3 is 19 g/day and ages 4–8 is 25 g/day. For boys, ages 9–13, the DRI recommendations are 31 g/day, and 38 g/days for ages 14–18. For girls ages 9–18, the DRI recommendations is 26 g/day and 25 g of dietary fiber is the recommended amount in a 2000-kcal diet [32, 33]. Manufacturers are allowed to call a food a "good source of fiber" if it contains 10% of the recommended amount (2.5 g/serving) and an "excellent source of fiber" if the food contains 20% of the recommended amount (5 g/serving). It is assumed that once the lay public knows of the benefits of a given functional food then they

	Recommended intake (g/MJ)		Recommended intake (g/d)	
Country	**Children**	**Adults**	**Children**	**Adults**
Belgium				F: 30 M 15–18 years: 40 M 19–≥75 years: 30
Estonia			5+ age	25–35
Europe		3		>25
France			5+ age	30
Germany/Austria		F:3		≥ 30
Switzerland		M:2.4		
Greece				EC 1993 recommendations
Hungary				25[1]
Ireland				25–35
Italy	Developmental age (≥ 1 year):2	3–4 (RI)		
Latvia				No recommendation
Lithuania				No recommendation
Luxembourg				No recommendation
The Netherlands	13 years: 2.8		3.4	32–45 g/d
	4–8 years: 3.0			
	9–13 years: 3.2			
	14–18 years: 3.4			
Nordic countries	1–17 years: 2–3, from school age (6 years) increasing to level recommended for adults	3	≥6 years (school age): 10,	F:25
			Increasing gradually to adult recommended level	M:35
Poland			1–3 years:0	19–30 years:25
			4–6 years: 19	31–50 years:25
			7–9 years: 19	51–65 years:25
			10–12 years:16	66–75 years:20
			13–15 years:19	>75 years:20
			16–18 years:21	
Portugal				27–40
Romania				25–35[2]
Solvakia			0–6 months: 1	F 19–54 years:22–26[3]
			7–12 months:3	M 19–34 months 26–32[3]
			1–3 years: 10	M 35–59 years: 24–30[3]

	Recommended intake (g/MJ)		Recommended intake (g/d)	
Country	Children	Adults	Children	Adults
			4–6 years: 14	M 35–39 years: 20
			7–10 years:17	M 60–74 years: 22
			F 11–14 years:18	F > 75 years: 18
			M 11–14 years: 20	M > 75 years: 20
			F 15–18 years: 18–22	Pregnant F: 26
			M 15–18 years: 22–25	Lactating F: 28
Solvenia	2.4	F:3		30
		M2.4		
UK				18 (range 12–24)[4]
USA	All ages ≥1 years: 3·4	All ages ≥1	1–3 years:19	F 19–50 years: 25
		Years: 3–4	4–8 years: 25	M 19–50 years: 38
			F 9–13 years: 26	F > 51 years:21
			M 9–13 years:31	M > 51 years: 30
			F 14–18 years:26	
			M 14–18 years: 38	
			1–3 years:14	F:25
Australia/New Zealand			4–8 years: 18	M: 30
			F 9–13 years: 20	
			M9–13 years: 24	
			F14–18 years: 22	
			M14–18 years: 28	
WHO				>25

F, female; M, male; RI, recommended intake.; 1-Insoluble fiber: soluble fiber ratio 3:1; 2–75% insoluble, 25% soluble; 3- Depending on physical activity; 4 NSP.
Source: Stephen et al. [23].

Table 2.
Recommendations (adequate intake) for average population total fiber intake in different age groups in Europe.

will embrace the dietary change. **Table 2** provides recommendations for adequate intake for dietary fiber in different countries for different age groups.

Despite the beneficial health effects of dietary fiber, its intake is far below recommendations globally. Although there is no known deficient state of dietary fiber reported worldwide, trends in recent studies of adherence to healthy lifestyle habits have strongly influence food choice, which in turn affect daily total fiber intake. This trend is common among the young adults, for their preference in the consumption of soft drinks, snacks, prepared and pre-cooked meals and other ready-to-eat products, most of which are rich in sugars and fats, and deficient in fiber [34, 35]. This has

Food sources	TDF(g/100 g)	Solube fiber (%TDF)	Insolube fiber (%TDF)	Cellulose	Pectic polysaccharide	Hemicellulose	β-Glucans	RS	RO
Vegetable (Excluding potatotes) Raw, steamed and baked vegetables	<0.5–6: median: 2.2	37	63	+ +	+ +	++ ++	- -	- -	- -
Soups	0.3–1.8; median: 0.9	39	6	+	+	++	—	—	—
Fruits									
Fresh fruit	0.4–10-4; median: 2.3	43	57	+ mostly skin	++	+	—	+(banana) (most fruits)	—
Processed fruit	0.4–2.0; median: 1.3			+	++	+	—	—	—
Dried fruit	0.1–11-14	53	47	+	++	+	—	—	—
Fruit Juices and nectar	Traces-0.65; median:0.4	90 (orange juice)	10	+	+++	—	—	—	—
Nuts and seeds	1.3–14; median:4.2	32	66	+	+	++	—	—	—
Legumes	4.2–10.6: median:4.5	25	75	+	—	++	—	++	++
Potatoes and other starchy tubers	0.5–8, median: 2.25	48 (potato without skin	52		+	+	—	+	—
Cereal products					—				
Leavened breads					+				—
White flour	3.0–3.4	50	50	+	—	++	—	+	—
Whole flour	5.6–7.2	20	80	+	—	++	+	+	+

Food sources	TDF(g/100 g)	Solube fiber (%TDF)	Insolube fiber (%TDF)	Cellulose	Pectic polysaccharide	Hemicellulose	β-Glucans	RS	RO
Breakfast cereals and cereal bars(excluding oat porridge, cereals)	1.2–15; median: 3–4	27	73	+	—	++	+	+	+
Oat porridge	1.7	52	48	+	—	++	++	+	+
Rye-based products	3.9–5-9	44	56	+	—	++	++	+	+
Rice									
White	0.82- < 1.1	≈0	≈100	+	+	+		+	
Whole	2.1–4; median 3–4	13	87	++	—	++		+	

TDF-total dietary fiber; RS- resistant starch; RO- resistant oligosaccharide; – means absent; + means present; ++ represent the level of abundance.
Source [23].

Table 3.
Dietary fiber in different food sources – Quantitative and qualitative aspects.

led to the creation of strategies, such as the formulation of dietary guidelines model to increase fiber intake to improve American people's health [24, 29, 35, 36]. Low fiber intake has also been reported for other countries like Tunisia [37], Spain [38] and Australia [39] where only 29.5% of women reported to meet the recommended Adequate Intake (AI) of dietary fiber during pregnancy of 28 g/day. In terms of tolerance level for fiber intake, there is no upper tolerable level, but it varies by individual, and overconsumption can lead to common side effects such as bloating and abdominal discomfort.

2.3 Dietary fiber food sources: fruits and vegetables for daily fiber target provision

Dietary fiber food sources are characterized with high total fiber, low in moisture and lipid content, low caloric value, and neutral flavor [40]. With dietary fiber being undoubtedly an essential part of a healthy diet, numerous studies in food science field and human nutrition have constantly encouraged that the public should include adequate amount of food rich in dietary fiber into the diet. Fruits, and vegetables, grains, legumes, nuts, bread, pasta, cereals, and seeds contribute significant amounts of fiber to the diet. These food groups account for more than 70% of dietary fiber in the food supply [41, 42]. Other fiber sources include over-the-counter laxatives containing fiber, fiber supplements, and fiber-fortified foods. Several surveys have reported the contribution from these major sources of dietary fiber in the diet [23, 29, 43]. However, the contribution of fiber proportion from each food source varies for different nations of the world. Although Some uniformity exists and hence some general statements, like "grain products provide the largest proportion of fiber in the diet" were reported in most reports, with bread being the largest grain source, with smaller contributions from cereals, pasta, biscuits, and pastries [29, 44]. White flour and potatoes have also been reported to be the major fiber sources in American diets because they are widely consumed in the United States [45]. **Table 3** gives information on the main dietary fiber food sources, components of dietary fiber found in different fiber food sources and their total fiber in gram per 100 g. Apart from nuts and seeds, dried fruits and vegetables are seen at the top of the chart with total dietary fiber of 0.1–11.4 and 0.5–8 g/100 g, respectively.

3. Fruit and vegetables for daily target provision

Fruits and vegetables have historically held a place in dietary guidance as most countries have dietary recommendations that include fruits and vegetables. Compared with grains products and other fiber sources, fruits and vegetables are a very good and cheap source of dietary fiber mainly because of their availability, relatively higher soluble/insoluble fiber ratio, higher fat retention capacity, enhanced colonic fermentability, and improved functionality [40–42]. The nutritional assessment of fruits and vegetables in terms of dietary fiber was calculated using The Harvard University food composition database, derived from the US Department of Agriculture (USDA) data, and were categorized as "high in fiber" [46]. Dietary guidelines for Americans (2010) recommend that one-half of the food plate should be fruit and vegetables because commonly consumed vegetables provide about 1 to 3 g dietary fiber per 100 g and supply vitamins and minerals to the diet. Additionally, fruit and vegetables are sources of phytochemicals that functions as phytoestrogens, anti-inflammatory agents, and other protective mechanisms [43].

Recent focus on fruit and vegetables as a promising source of dietary fiber resides not only in their important role of lowering the risk of chronic diseases, but also as rich sources of carbohydrates, fats, proteins, energy, vitamins A, B1, B2, B3, B6, B9, B12, C, folic acid, and minerals [45]. The energy and fiber content (soluble and insoluble) of 10 commonly consumed fruits and vegetables are presented in **Table 4**. Most fruits and vegetables are not only consumed raw, some are consumed in the cooked form, and sometimes fried or prepared with other ingredients prior to consumption form. Potato is known to be one of the nutrient dense vegetable with high fiber content. Cooking methods, including frying, do not diminish dietary fiber content of potato. Fried and oven-baked potato with skin is reported to have fiber concentration ranging from 0.6 to 5.1 and may contribute a great amount of sodium and fat to the diet (**Table 3**). Similarly, cooked broccoli, green beans, spinach, and corn contains 3.3 g, 3.2 g, 2.4 g, 2.4 g respectively [45]. Regardless of the variation in the nutritional composition of fruits and vegetables varies widely, but they are known to be good sources of fiber and potassium [47].

Several studies have endorsed fruits and vegetables as a suitable and healthy source for adequate fiber intake. World Health Survey (2002–2003) report gotten from the interviews of 196,373 adults from 52 countries with mainly small and middle income showed that about 78% of the men and women consumed ≤5 portions

Common fruit /vegetable	Serving	Kcal	Total dietary fiber	Insoluble dietary fiber	Soluble dietary fiber
Potato, boiled	1 med, 167 g	144	3.0	1.6	1.4
Iceberg lettuce	1 cup, 57 g	8	0.7	0.6	0.1
Tomato	NLEA, 148 g	27	1.8	1.6	0.2
Onion	NLEA, 148 g	47	1.3	0.8	0.5
Carrot	NLEA, 85 g	30	2.5	2.1	0.4
Celery	NLEA, 110 g	18	1.8	1.7	0.1
Sweet corn	1 ear, 77 g	74	1.8	1.7	0.1
Broccoli	NLEA, 148 g	50	3.8	3.0	0.8
Green cabbage	1 cup, 89 g	22	2.2	1.7	0.1
Cucumber with peel	1 cup, slices	16	0.6	0.5	0.1
Banana	1 med, 118 g	105	3.1	2.1	1.0
Apple with skin	1 med, 182 g	95	4.4	3.1	1.3
Watermelon	NLEA, 280 g	84	1.1	0.8	0.3
Orange	NLEA, 154 g	75	3.4	1.4	2.0
Cantaloupe	NLEA, 134 g	46	1.2	0.9	0.3
Green grapes	NLEA, 126 g	87	1.1	0.6	0.5
Grapefruit	NLEA, 154 g	65	2.5	0.9	1.6
Strawberry	NLEA 147 g,	47	2.9	2.2	0.7
Peach	NLEA, 147 g	57	2.2	1.2	1.0
Pear	NLEA, 166 g	96	5.1	3.6	1.5

Source: [43].

Table 4.
Ten commonly consumed fruits and vegetables in.

of vegetables and fruit daily as recommended by the World Health Organization (WHO, according to the WHO: 400 g/day) [48]. In a 24-year study, Bertoia et al. [46] found that the consumption of all fruits and most vegetables contributed substantially to meet the daily adequate fiber intake by combining an increase of one-to-two servings of vegetables and one-to-two servings of fruits daily.

To meet the recommended (DRI) daily allowance of 25 to 38 g/day, Hornick and Weiss [49] found that a diet with mean fruits and vegetables of 2 daily servings intakes (servings/d) of 5.16 servings 3.5 portions (men); and 4.7 servings, 3.8 portions (women) will help an individual to meet the daily target. About 10% of the western population have been found to achieve the recommended adequate fiber intake by consuming whole fruits like apples/pears and prunes during mealtimes [50]. Fruits and vegetables are modest for synthetizing functional fiber. By-products from processing of fruits and vegetables like orange, apple and peach, carrot, potato, green pea, pepper, artichoke, onion, and asparagus that contains both soluble and insoluble fiber compounds have been reported to have a great potential for enrichment of foods or for providing techno-functional properties to food [51].

Likewise, dietary fiber content of fruits and vegetables have been implicated with several health-promoting properties. Prentice and Jebb [52] reported dietary fiber from fruits and vegetables to be involved in the regulation of hunger and saturation and hence prevented obesity. In a meta-analysis of cohort studies, Kan et al. [53] showed that dietary fiber from fruits and vegetables reduce the risk of diabetes. Dietary fiber from Fruits was also found to be associated with a reduced risk of chronic obstructive pulmonary disease [54]. Boeing et al. [55] reported that the dietary approaches to stop hypertension consists of a high proportion of vegetables and fruit in the diet. Some types of dietary fiber which are not present in most fiber food sources, are found in some fruits. For example, resistant starch that is only present in starchy foods is also present in fruits like green banana. Likewise, pectic substances are found in most fruits and vegetables but absent in major fiber sources (**Table 3**). The Los Angeles Atherosclerosis Study observed that a higher intake of pectin significantly slowed intima-media thickness IMT progression which appears to be the leading cause of cardiovascular disease [56].

It is a known fact that dietary fibers obtained from different sources, vary in fiber content and in composition of other nutrients that may impact their health benefits. For example, fiber from different food sources behave differently during their transit through the gastrointestinal tract, depending on their chemical composition and physicochemical characteristics [33, 57, 58]. Higher intake of dietary fibers from fruits like red apples, pears and prunes were reported to be associated with a lower the risk of diverticulitis [59]. Recent research found that dietary fiber from white potatoes plays a role in the production of fecal short-chain fatty acids concentration, which is important for immune regulation and maintaining gut health [45]. Studies have also established that most vegetable fiber, especially potato fiber has antiproliferative functions that may act as chemo preventive agents and protect the small intestinal wall against ingested compounds formed during cooking, such as melanoidins and acrylamide [60].

3.1 Health promoting properties of dietary fiber

The physiological functionalities and health benefits of dietary fiber has been extensively studied. Deficiency of dietary fiber in the diet may lead to several diseases such as gastrointestinal disorder [58], metabolic syndrome [61], appendicitis [62], inflammatory bowel syndrome [63], diabetes [64], obesity [65], cardiovascular disease [24], gallstones [66], etc. **Table 5** summarizes the functional health effect of dietary fiber.

Source	Isolated dietary fiber	Functions	Effects
Potato, lentils, peas	Somalto-oligosaccharides	Reduction of LDL and total cholesterol	Reduced risk of heart disease and cholesterolaemia
Apples, carrots, oranges, grapefruits, and lemons	Pectin	Blood lipid lowering; attenuates blood glucose	May reduce onset risk or symptoms of metabolic syndrome and diabetes
Treated vegetable cellulose	Methylcellulose,	Adds bulk to stool	Alleviates constipation.
artichokes, asparagus, bananas, chicory root, garlic, onions	Inulin, fructooligosacchaides	Balances intestinal pH and stimulates intestinal fermentation. production of short-chain fatty acids	May reduce the risk of colorectal cancer
Konjac roots	Glucomannan	Speeds the passage of foods through the digestive system	Facilitates regularity
Corn	Resistant starch, beta-glucan	Adds bulk to the diet, making feel full faster	May reduce appetite
Apples, oats, and Barley	Psyllium, β-glucans, and pectin	Increases gut transit rate and reduces the amount of time available for colonic bacterial fermentation of non-digested foodstuff	Lowers variance in blood sugar levels
Green banana	Cellulose	Delayed colonic transit time	Prolonged post-meal satiety and satiation
Beans		Induction of cholecystokinin	Balances intestinal pH and stimulates intestinal fermentation
Psyllium husk	Psyllium	Reduction of the glycaemic response	Lower the risk of diabetes

Sources: [19, 51, 67].

Table 5.
Source, types of fiber, functions, and health benefits of dietary fiber.

4. Conclusion

Dietary fiber is an active component of fruits and vegetables. They have been recognized nutritionists and public health as the cheapest and most commonly available fiber source all over the world. Apart from their dietary fiber composition, fruits and vegetables contain nutrients such as vitamins, minerals, and phytochemicals, including polyphenols, which provide support for their biological plausibility and enhance their health benefits. Increase in the consumption of fruit and vegetable fiber increases post-meal satiety and a decrease in subsequent hunger. Addition of fruits and vegetables servings to the diet can provide the daily recommended adequate fiber intake, therefore public health recommendations and nutritional guidelines should emphasize adequate consumption of specific fruits and vegetables and provide practical dietary guidance that maximizes the potentials their dietary fiber content and composition.

Acknowledgements

Authors acknowledge the financial support of Govan Mbeki Research Development Centre, University of Fort Hare. Grant number: C127.

Conflict of interest

The authors declare no conflict of interest.

Author details

Oluwafunmilayo Dorcas Adegbaju*, Gloria Aderonke Otunola and Anthony Jide Afolayan
University of Fort Hare, Alice, South Africa

*Address all correspondence to: oadegbaju@ufh.ac.za

References

[1] Zimmet PZ, Magliano DJ, Herman WH, Shaw JE. Diabetes: a 21st century challenge. The lancet Diabetes & endocrinology. 2014; 2: 56-64. doi: 10.1016/S2213-8587(13)70112-8.

[2] Ortega FB, Lavie CJ, Blair SN. Obesity and cardiovascular disease. Circulation research. 2016; 118: 1752-1770. doi: 10.1161/CIRCRESAHA.115.306883.

[3] Bennett CM, Coleman HG, Veal PG, Cantwell MM, Lau, CC, Murray LJ. Lifestyle factors and small intestine adenocarcinoma risk: A systematic review and meta-analysis. Cancer epidemiology. 2015; 39: 265-273. doi: 10.1016/j.canep.2015.02.001.

[4] Barber TM, Kabisch S, Pfeiffer AF, Weickert MO. The health benefits of dietary fiber. Nutrients. 2020; 12: 3209. doi: 10.3390/nu12103209

[5] Sarwer DB, Polonsky HM. The psychosocial burden of obesity. Endocrinol. Metab. Clin. N. Am. 2016; 45:677-688. doi: 10.1016/j.ecl.2016.04.016.

[6] Nena E, Steiropoulos P, Constantinidis TC, Perantoni E, Tsara V. Work productivity in obstructive sleep apnea patients. Journal of occupational and environmental medicine. 2010; 52(6), 622-625. doi: 10.1097/JOM.0b013e3181e12b05.

[7] Tremmel M, Gerdtham UG, Nilsson PM, Saha, S. Economic burden of obesity: a systematic literature review. International journal of environmental research and public health.2012; 14:435. doi: 10.3390/ijerph14040435.

[8] Birger M, Kaldjian AS, Roth GA, Moran AE, Dieleman JL, Bellows BK. Spending on Cardiovascular Disease and Cardiovascular Risk Factors in the United States: 1996-2016. Circulation. 2021: April 30.doi: 10.1161/CIRCULATIONAHA.120.053216

[9] Yapo BM, Koffi KL. Dietary Fiber Components in Yellow Passion Fruit Rind: A Potential Fiber Source. Journal of Agricultural and Food Chemistry. 2008;56: 5880-5883. doi: 10.1021/jf073247p

[10] Jimoh MO, Afolayan AJ, Lewu FB. Suitability of Amaranthus species for alleviating human dietary deficiencies. South African Journal of Botany. 2018; 115: 65-73. doi.org/10.1016/j.sajb.2018.01.004

[11] Institute of Medicine 2006. Dietary Reference Intakes: The Essential Guide to Nutrient Requirements. The National Academies Press. 2006. Washington, DC: https://doi.org/10.17226/11537

[12] Hipsley EH. Dietary "fibre" and pregnancy toxaemia. British medical journal. 1953 Aug 22;2(4833):420.

[13] Codex Alimentarius Commission. 2010. [online], http:// www.codexalimentarius. net/download/standards/34/CXG _002e.pdf [accessed: 15 July 2021]

[14] Trowell H, Southgate DT, Wolever TS, Leeds A, Gassull M, Jenkins DA. Dietary fiber redefined. The Lancet. 1976 May 1;307(7966):967. doi: 10.1016/s0140-6736(76)92750-1.

[15] Southgate DA. Definitions and terminology of dietary fiber. In Dietary fiber in health and disease 1982 (pp. 1-7). Springer, Boston: MA; 1982. p.1-7. DOI: 10.1007/978-1-4615-6850-6_1

[16] Hijová E, Bertková I, Štofilová J. Dietary fiber as prebiotics in nutrition. Central European journal of public health. 2019; Sep 1;27:251-5. doi: 10.21101/cejph.a5313.

[17] Kohn JB. Is dietary fiber considered an essential nutrient?. Journal of the Academy of Nutrition and Dietetics. 2016; Feb 1;116(2):360. doi.org/10.1016/j.jand.2015.12.004

[18] Knudsen KB. The nutritional significance of "dietary fiber" analysis. Animal feed science and technology. 2001; Mar 15;90(1-2):3-20. doi.org/10.1016/S0377-8401(01)00193-6

[19] Dhingra D, Michael M, Rajput H, Patil RT. Dietary fiber in foods: a review. Journal of food science and technology. 2012; Jun; 49:255-66. doi: 10.1007/s13197-011-0365-5

[20] Slavin J. Impact of the proposed definition of dietary fiber on nutrient databases. Journal of Food Composition and Analysis. 2003 Jun 1;16(3):287-91. doi:10.1016/S0889-1575(03)00053-X

[21] Gropper SS, Smith JL. Advanced nutrition and human metabolism. Cengage Learning; 2012 Jun 1.

[22] Chau CF, Huang YL. Effects of the insoluble fiber derived from *Passiflora edulis* seed on plasma and hepatic lipids and fecal output. Molecular nutrition & food research. 2005; Aug;49(8): 786-90. doi.org/10.1002/mnfr.200500060

[23] Stephen AM, Champ MM, Cloran SJ, Fleith M, van Lieshout L, Mejborn H, Burley VJ. Dietary fiber in Europe: current state of knowledge on definitions, sources, recommendations, intakes and relationships to health. Nutrition research reviews. 2017 Dec;30(2):149-90. doi:10.1017/S095442241700004X

[24] Soliman GA. Dietary fiber, atherosclerosis, and cardiovascular disease. Nutrients. 2019; May;11(5): 1155. doi.org/10.3390/nu11051155

[25] Ma W, Nguyen LH, Song M, Jovani M, Liu PH, Cao Y, Tam I, Wu K, Giovannucci EL, Strate LL, Chan AT. Intake of dietary fiber, fruits, and vegetables, and risk of diverticulitis. The American journal of gastro enterology. 2019 Sep;114(9):1531. doi: 10.14309/ajg.0000000000000363

[26] U.S. Department of Agriculture and U.S. Department of Health and Human Services. Dietary Guidelines for Americans, 2010. 7th Edition, Washington, DC: U.S. Government Printing Office, December, 2010.

[27] U.S. Department of Health and Human Services and U.S. Department of Agriculture Dietary Guidelines for Americans 2015-2020. [(accessed on 4 August 2018)]; Available online: http://health.gov/dietaryguidelines/ 2015/guideline

[28] You A. Dietary guidelines for Americans. US Department of Health and Human Services and US Department of Agriculture. 2015.

[29] Hoy MK, Goldman JD. Dietary fiber intake of the US population, what we eat in America, NHANES 2009-2010. US Department of Agriculture. 2014. Oct 28;12(6).

[30] World Health Organization. Global report on diabetes: executive summary. World Health Organization; 2016.

[31] Fayet-Moore F, Cassettari T, Tuck K, McConnell A, Petocz P. Dietary fibre intake in Australia. Paper I: associations with demographic, socio-economic, and anthropometric factors. Nutrients. 2018 May;10(5):599. doi.org/10.3390/nu10050599

[32] Lupton JR, Brooks JA, Butte NF, Caballero B, Flatt JP, Fried SK. Dietary reference intakes for energy, carbohydrate, fiber, fat, fatty acids, cholesterol, protein, and amino acids. National Academy Press: Washington, DC, USA. 2002;5:589-768. Institute of

Medicine, Food and Nutrition Board Web site. http://www.iom.edu/Reports/

[33] King DE, Mainous III AG, Lambourne CA. Trends in dietary fiber intake in the United States, 1999-2008. Journal of the Academy of Nutrition and Dietetics. 2012 May 1;112(5):642-8. doi: 10.1016/j.jand.2012.01.019

[34] Chourdakis M, Tzellos T, Papazisis G, Toulis K, Kouvelas D. Eating habits, health attitudes and obesity indices among medical students in northern Greece. Appetite. 2010 Dec 1;55(3):722-5.

[35] Myhre JB, Løken EB, Wandel M, Andersen LF. The contribution of snacks to dietary intake and their association with eating location among Norwegian adults–results from a cross-sectional dietary survey. BMC Public Health. 2015 Dec;15(1):1-9. doi: 10.1186/s12889-015-1712-7

[36] McGill CR, Devareddy L. Ten-year trends in fiber and whole grain intakes and food sources for the United States population: National Health and Nutrition Examination Survey 2001-2010. Nutrients. 2015 Feb;7(2):1119-30. DOI: 10.3390/nu7021119

[37] Mokhtar N, Elati J, Chabir R, Bour A, Elkari K, Schlossman NP, Caballero B, Aguenaou H. Diet culture and obesity in northern Africa. The Journal of nutrition. 2001 Mar;131(3): 887S-92S. doi: 10.1093/jn/131.3.887S.

[38] Navarro AA, Palomar JC, Hernández-Agero TO, Oliver AP, Ortiz JC, Conde MF, Jubete FF, de los Mozos Pascual M, Vioque RS, Cobos PD, Rodríguez RL. Informe del Comité Científico de la Agencia Española de Seguridad Alimentaria y Nutrición (AESAN) sobre el consumo humano ocasional de almortas (*Lathyrus sativus*). Revista del Comité Científico de la AESAN. 2010(11):9-20.

[39] Pretorius RA, Palmer DJ. High-Fiber Diet during Pregnancy Characterized by More Fruit and Vegetable Consumption. Nutrients. 2021 Jan;13(1):35. DOI: 10.3390/nu13010035

[40] Larrauri JA. New approaches in the preparation of high dietary fiber powders from fruit by-products. Trends in Food Science & Technology. 1999 Jan 1;10(1):3-8.

[41] Marlett JA, Cheung TF. Database and quick methods of assessing typical dietary fiber intakes using data for 228 commonly consumed foods. Journal of the American Dietetic Association. 1997 Oct 1;97(10):1139-51.

[42] Amaya-Cruz, DM., Rodríguez-González, S. Pérez-Ramírez, IF,, Loarca-Piña, G., Amaya-Llano, S., Gallegos-Corona, MA., Reynoso-Camacho, R. Juice by-products as a source of dietary fiber and antioxidants and their effect on hepatic steatosis. Journal of Functional Foods, 2015. 17, 93-102. https://doi.org/10.1016/j.jff.2015.04.051

[43] Slavin JL, Lloyd B. Health benefits of fruits and vegetables. Advances in nutrition. 2012 Jul;3(4):506-16. doi:10.3945/an.112.002154

[44] Joye IJ. Dietary fibre from whole grains and their benefits on metabolic health. Nutrients. 2020 Oct;12(10):3045. doi: 10.3390/nu12103045

[45] Storey M, Anderson P. Income and race/ethnicity influence dietary fiber intake and vegetable consumption. Nutrition Research. 2014 Oct 1;34(10):844-50. doi.org/10.1016/j.nutres.2014.08.016

[46] Bertoia ML, Mukamal KJ, Cahill LE, Hou T, Ludwig DS, Mozaffarian D, Willett WC, Hu FB, Rimm EB. Changes in intake of fruits and vegetables and weight change in United States men and women followed for up to 24 years: analysis from three prospective cohort studies. PLoS medicine. 2015 Sep

22;12(9):e1001878. DOI:10.1371/journal.pmed.1001878

[47] Figuerola, F., Hurtado, ML., Estévez, AM., Chiffelle, I. Asenjo, F. Fiber concentrates from apple pomace and citrus peel as potential fiber sources for food enrichment. Food Chemistry. 2005. 91(3), 395-401.

[48] Hall JN, Moore S, Harper SB, Lynch JW. Global variability in fruit and vegetable consumption. American journal of preventive medicine. 2009 May 1;36(5):402-9. doi:10.1016/j.amepre.2009.01.029

[49] Hornick BA, Weiss L. Comparative nutrient analysis of commonly consumed vegetables: support for recommending a nutrition education approach emphasizing specific vegetables to improve nutrient intakes. Nutrition today. 2011 May 1;46(3):130-7.

[50] Dreher ML. Whole fruits and fruit fiber emerging health effects. Nutrients. 2018 Dec;10(12):1833. doi: 10.3390/nu10121833

[51] Li YO, Komarek AR. Dietary fibre basics: Health, nutrition, analysis, and applications. Food quality and safety. 2017 Mar 1;1(1):47-59. doi:10.1093/fqs/fyx007 Review doi:10.1093/fqs/fyx007

[52] Prentice AM, Jebb SA. Fast foods, energy density and obesity: a possible mechanistic link. Obesity reviews. 2003 Nov;4(4):187-94. doi: 10.1046/j.1467-789x.2003.00117.x

[53] Kan H, Stevens J, Heiss G, Rose KM, London SJ. Dietary fiber, lung function, and chronic obstructive pulmonary disease in the atherosclerosis risk in communities study. American journal of epidemiology. 2008 Mar 1;167(5):570-8. doi: 10.1093/aje/kwm343

[54] Varraso R, Willett WC, Camargo Jr CA. Prospective study of dietary fiber and risk of chronic obstructive pulmonary disease among US women and men. American journal of epidemiology. 2010 Apr 1;171(7):776-84. doi: 10.1093/aje/kwp455

[55] Boeing H, Bechthold A, Bub A, Ellinger S, Haller D, Kroke A, Leschik-Bonnet E, Müller MJ, Oberritter H, Schulze M, Stehle P. Critical review: vegetables and fruit in the prevention of chronic diseases. European journal of nutrition. 2012 Sep;51(6):637-63. DOI 10.1007/s00394-012-0380-y

[56] Wu H, Dwyer KM, Fan Z, Shircore A, Fan J, Dwyer JH. Dietary fiber and progression of atherosclerosis: the Los Angeles Atherosclerosis Study. The American journal of clinical nutrition. 2003 Dec 1;78(6):1085-91. https://academic.oup.com/ajcn/article-abstract/78/6/1085/4677517

[57] Carlson JL, Erickson JM, Lloyd BB, Slavin JL. Health effects and sources of prebiotic dietary fiber. Current developments in nutrition. 2018 Mar 1;2(3): doi: 10.1093/cdn/nzy005.

[58] Capuano E. The behavior of dietary fiber in the gastrointestinal tract determines its physiological effect. Critical reviews in food science and nutrition. 2017 Nov 2;57(16):3543-64. doi: 10.1080/10408398.2016.1180501

[59] Ma W, Nguyen LH, Song M, Jovani M, Liu PH, Cao Y, Tam I, Wu K, Giovannucci EL, Strate LL, Chan AT. Intake of dietary fiber, fruits, and vegetables, and risk of diverticulitis. The American journal of gastro enterology. 2019 Sep;114(9):1531. doi.org/10.14309/ajg.00000000 00000363

[60] Dobrowolski P, Huet P, Karlsson P, Eriksson S, Tomaszewska E, Gawron A, Pierzynowski SG. Potato fiber protects the small intestinal wall against the toxic influence of acrylamide. Nutrition. 2012

Apr 1;28(4):428-35.doi.org/10.1016/j.nut.2011.10.002

[61] Chen JP, Chen GC, Wang XP, Qin L, Bai Y. Dietary fiber and metabolic syndrome: a meta-analysis and review of related mechanisms. Nutrients. 2018 Jan;10(1):24. doi: 10.3390/nu10010024

[62] Damanik B, Fikri E, Nasution IP. Relation between Fiber Diet and Appendicitis Incidence in Children at H. Adam Malik Central Hospital, Medan, North Sumatra-Indonesia. DOI:10.15562/bmj.v5i2.225

[63] El-Salhy M, Ystad SO, Mazzawi T, Gundersen D. Dietary fiber in irritable bowel syndrome. International journal of molecular medicine. 2017 Sep 1;40(3):607-13. DOI: 10.3892/ijmm.2017.3072

[64] Weickert MO, Pfeiffer AF. Impact of dietary fiber consumption on insulin resistance and the prevention of type 2 diabetes. The Journal of nutrition. 2018 Jan 1;148(1):7-12. rom https://academic.oup.com/jn/article-abstract/148/1/7/4823705

[65] Delzenne NM, Olivares M, Neyrinck AM, Beaumont M, Kjølbæk L, Larsen TM, Benítez-Páez A, Romaní-Pérez M, Garcia-Campayo V, Bosscher D, Sanz Y. Nutritional interest of dietary fiber and prebiotics in obesity: Lessons from the MyNewGut consortium. Clinical Nutrition. 2020 Feb 1;39(2):414-24. (http://creativecommons.org/licenses/by-nc-nd/4.0/

[66] Zhang JW, Xiong JP, Xu WY, Sang XT, Huang HC, Bian J, Xu YY, Lu X, Zhao HT. Fruits and vegetables consumption and the risk of gallstone diasease: A systematic review and meta-analysis. Medicine. 2019 Jul;98(28).doi.org/10.1097/MD.0000000000016404

[67] Holscher HD. Dietary fiber and prebiotics and the gastrointestinal microbiota. Gut microbes. 2017 Mar 4;8(2):172-84. 10.1080/19490976.2017.1290756

Chapter 5

Prebiotic Dietary Fibers for Weight Management

Ceren Gezer and Gözde Okburan

Abstract

While all prebiotics are accepted as dietary fibers, not all dietary fibers are accepted as prebiotics. Fructo-oligosaccharides and galacto-oligosaccharides are significant prebiotic dietary fibers related with the regulation of weight management. They, selectively stimulate the growth of *bifidobacteria* and *lactobacillus*, thus help to modulate gut microbiota. Since *bifiodobacteria* population are responsible for energy scavenging they are playing a vital role in the weight management. In addition, prebiotics fermented to short chain fatty acids by gut microbiota, whose presence in the large intestine is responsible for many of the metabolic effects and prevent metabolic diseases such as obesity. Short chain fatty acids via different mechanisms also stimulate satiety hormones such as GLP-1 and PYY, and shift glucose and lipid metabolism. To conclude, prebiotic dietary fibers beneficially impact the gut microbiota thus can be effective on regulation of weight management. There is a need for further clinical trials to explain more comprehensively the effects of dietary prebiotics on weight management.

Keywords: prebiotics, dietary fiber, obesity, weight management, short chain fatty acids, gut microbiota

1. Introduction

Over the few past decades, the prevalence of obesity has been seen to increase dramatically and thus this increase lead the attention towards environment. As a result of socioeconomic development, especially in Western countries, increased sedentariness, nearby abundant presence of cheap high-energy dense foods have all been concerned as significant contributing factors [1]. In this context, the modified fatty acid composition of a Western diet, often rich in saturated and trans fatty acids, raises serum total and low density lipoprotein (LDL) cholesterol levels, increasing the risk of chronic vascular diseases. Furthermore, diets high in sodium and low in potassium can lead to a variety of chronic diseases, including hypertension and stroke. Another factor is the presence of dietary fibers such as inulin, resistant starch and beta-glucan, which are important food components that are reduced in the Western diet and can delay gastric emptying, reduce appetite and therefore help regulate dietary energy intake [2].

The dietary fiber firstly defined in 1950s as non-digestible components of plant cell wall, then in late 1970s dietary fiber defined as polysaccharides and lignin which are resistant to enzymatic digestion [3]. So, the definitions are mainly focusing on non-digestibility. During years, dietary fiber description and assessment methods has been improved. In 2001, in addition to the non-digestibility,

American Association of Cereal Chemists defined dietary fiber as which is also show partial fermentation in colon. In 2008, the European Commission defined dietary fiber as carbohydrate polymers have the degree of polymerization three or more monomeric units and not digest in small intestine by enzymes. In 2009 Codex Alimentariius also defined dietary fiber as carbohydrate polymers with the degree of polymerization ten or more monomeric units which not digest in small intestine by enzymes [4]. Fiber can be classified according to chemical composition, solubility and fermentability. According to chemical composition, dietary fiber can be classified as non-starch polysaccharides, resistant oligosaccharides, resistant starch, lignin and other substances such as saponins, tannins, etc. [5]. Related with their chemical composition, solubility and fermentability of each dietary fiber type has been shown different physiological effects such as lowering gastric emptying and blood cholesterol level, production of short-chain fatty acids by fermentation. Thus, dietary fiber has been demonstrated potential prevention from chronic diseases such as cancer, cardiovascular diseases, diabetes and obesity [4]. Some fibers which fermented in colon are classified as prebiotics. Prebiotics has been defined as "substrate that is selectively utilized by host microorganisms conferring a health benefit" by International Scientific Association for Probiotics and Prebiotics in 2017 [6]. The substances has to be demonstrate three main properties to be defined as prebiotics. Firstly, it has to be resistant to digestive enzymes and gastric acidity; secondly, it can be fermented by gut microbiota and thirdly, it can be selectively stimulate the growth of beneficial gut bacteria such as *lactobacilli* and *bifidobacteria*. Therefore, while all prebiotics have been classified as dietary fiber, not all dietary fiber has been shown prebiotic properties [6, 7]. Prebiotic dietary fibers fermented to short chain fatty acids (SCFAs) such as acetate, propionate and butyrate by gut microbiota. SCFAs are responsible for many health effects by increasing *lactobacilli* and *bifidobacteria*, decreasing pathogenic bacteria, producing beneficial metabolites, decreasing protein fermentation, modulating gut barrier permeability and supporting immune system defense [8]. Thus, prebiotic dietary fibers alter gut microbiota positively to prevent metabolic diseases such as irritable bowel syndrome and crohn's disease, colorectal cancer, cardiovascular diseases and obesity. Fructo-oligosaccharides (FOSs) and galacto-oligosaccharides (GOSs) are well known prebiotic fibers which are also related with the regulation of weight management [9].

2. Dietary fibers with prebiotic properties

Fermentation of prebiotic dietary fibers by gut microbiota is related with solubility, longevity of chain and structure. Soluble fibers fermented better than insoluble fibers and oligosaccharides fermented better than polysaccharides. FOSs and GOSs are most common and proven prebiotic dietary fibers that meet the three main prebiotic properties which are resistance to digestive enzymes and acidity, fermentation by gut microbiota and selectively stimulation of the growth of *lactobacilli* and *bifidobacteria*. There is not sufficient randomized controlled clinical studies on other dietary fibers such as resistant starch 2, 3 and 4, pectins, polydextrose manno-oligosaccharides and xylo-oligosaccharides to proven them as prebiotic dietary fibers. However, there are *in vitro* and preclinical studies indicated that these fibers can be accepted as prebiotic dietary fiber candidate. While resistant starch 1 and 5, lignin and cellulose not recognized as prebiotic dietary fibers since it has not been proven that they meet the prebiotic properties [10].

Fructans are fructose polymers which has three main types, and inulin type fructans such as inulin, fructo-oligosaccharide and oligofructose are important ones. They contain β (2–1) linear chain of fructose. Degree of polymerization of fructooligosaccharide is less than 10 while inulin's polymerization degree is between 3 and 60. The degree of polymerization of oligofructose is between 2 and 20 and forming as a product of degradation of inulin [9, 11, 12]. FOSs are naturally found in asparagus, sugar beet, garlic, chicory, onion, Jerusalem artichoke, wheat, honey, banana, barley, tomato, and rye. It has been known that FOSs stimulate bifidogenic activity [13] GOSs are galactose polymers have β (1–3) and β (1–4) bonds and prebiotic GOSs have glucose as a terminal end and generally consist of 2–10 galactose and 1 glucose synthesized by β-galactosidase. GOSs are naturally found in milk. It has have been known that GOSs are stimulate bifidogenic activity [12, 13]. Therefore, FOSs and GOSs increase *bifidobacteria* [14] hence can switch glucose metabolism and show beneficial metabolic effects to control chronic diseases [15].

3. Effects of prebiotic dietary fiber in control of chronic diseases

There is an association of dietary fiber consumption with a healthy gut microbiome, also there are promising evidence which demonstrates a favorable effect of dietary fiber on body weight and overall metabolic health by reducing the risk for the development of cardiovascular disease and mortality. Moreover, there has been additional health benefits of dietary fiber such as reduction the risk of malignancy and improved colonic health [16, 17]. It has been supported by Academy of Nutrition and Dietetics in 2015 that the consumption of sufficient amount of dietary fiber reduces the risk of some chronic diseases such as diabetes, obesity and coronary heart diseases [18]. Precisely, studies have revealed that individuals with adequate intake of dietary fiber seems to be at lower risk for developing stroke, colorectal cancer, cardiovascular diseases and type-2 diabetes [19–26] Sufficient intake of dietary fiber is correspondingly related with lower blood pressure and lower serum cholesterol levels [27]. Additionally, via different mechanism through satiety or fullness regulation, adequate intake of dietary fiber is proposed to help in weight loss and weight management [28–33]. Furthermore, dietary fiber appears to improve immune function via gut health and fiber-microbiota interactions [34–36]. Increased consumption of high-fat and high-sugar diets have been shown to alter microbial ecology, leading to the impression that the gut microbiota may function as an environmental factor resulting in increased energy harvesting and obesity [2].

Numerous classes of prebiotic dietary fibers display diverse health benefits. FOSs and GOSs have long been considered prebiotics. Nonetheless, apart from those prebiotic dietary fibers other categories (guar, lactulose, maltodextrin, etc.) propose great health benefits, although in varying ranges of efficacies [8]. It has been indicated that GOSs, FOSs and inulin alter glucose and lipid metabolism hence can reduce body weight and the risk of chronic diseases such as diabetes and cardiovascular diseases [37]. A study conducted with overweight or obese individuals indicated that especially inulin-type fructans and GOSs have beneficial properties on metabolic endotoxemia. After they get fermented in the gastrointestinal tract, especially in general inulin-type fructans and in particular FOSs produce SCFAs, thus those fermentation products favor the development of beneficial microorganisms to the detriment of other harmful population. Likewise, those SCFAs significantly increases feelings of satiety and reduces feelings of hunger and thus reduces energy intake [38]. Although it has been known for years that the metabolic benefits of dietary fiber have positive effects on human health including the prevention of diabetes and obesity, the mechanism of these beneficial effects has not been fully defined until recent years. Thus far, SCFAs and their receptors have been progressively appreciated as a fundamental mediator that relates diet and the gut microbiota to host physiology by modulating endocrine responses, development and functioning of leukocytes, and the activity of enzymes and transcription factors. Consequently, it is essential to further lighten the role of SCFA receptors concerning to the efficacy of dietary interventions and gut microbiota manipulations in the management of obesity and metabolic syndrome [39].

4. Application of prebiotic dietary fiber on weight management

Manipulation of the gut microbiome, which is mainly influenced by diet, seems to be an innovative therapeutic tool to prevent or control obesity and related diseases. Of specific concern, prebiotic dietary fiber are fermented by the gut microbiota, which consequently providing potentially beneficial health effects [6]. Based on numerous studies which were conducted in animals and humans have been suggested that fermentable prebiotic dietary fibers may increase satiety, enhance obesity-related metabolic disorders, and modulate gut-related immunity [40–42]. Suggested mechanisms to explain such effects commonly comprise bacterial metabolites such as SCFAs. Inulin-type fructans are prebiotics that support *bifidobacteria* and produces SCFAs upon fermentation; their administration can improve health outcomes, particularly in the context of obesity [2, 6].

The effects of dietary fiber on weight management may be due to the non-digestible nature of dietary fibers, which prolongs transit times in the intestinal lumen and accordingly provides greater satiety compared to simple and easily digestible polysaccharides. Moreover, dietary fiber may also play a role in prolongation of meal intervals and cause an enhanced mastication on satiety via presumable cephalic and peripheral effects. Remarkably, diets rich in dietary fiber have lesser energy bulk and could affect the flavor and palatability of foods, which can eventually lead to lower energy intake [43]. Another mechanism for the appetite-reducing effects of dietary fiber is the stimulation of glucose-dependent insulinotropic peptide (GIP) signaling by gastrointestinal satiety peptide hormones such as glucagon-like peptide (GLP-1) and peptide YY (PYY) or dietary fiber resulting from their fermentation in the large intestine by the gut microbiota [44].

Other possible mechanism which relates the mainly prebiotic dietary fiber and weight regulation is fermented products of dietary fiber which are known as SCFAs. In humans, nutrient digestion and absorption occurs mainly in the stomach and proximal small intestine. Vital source of energy for the human being is the carbohydrates, but the ability of humans to break down and use dietary mono-, oligo- or polysaccharides is very limited. Various members of the gut microbiota, known as saccharolytic microorganisms, degrade these complex glycan's and in this manner providing the host with a variety of metabolites, particularly SCFAs. SCFAs which are the products of the digestion of soluble plant polysaccharides are not only an important source of energy, but also play key roles in the regulation of food and energy intake [45]. As mentioned previously, SCFAs and their receptors have been progressively appreciated as a fundamental mediator that relates diet and the gut microbiota to host physiology by modulating endocrine responses, development and functioning of leukocytes, and the activity of enzymes and transcription factors. Free fatty acid receptor (FFAR) 3/ G protein coupled receptor (GPR) 41 and FFAR2/GPR43 represent two SCFA-specific GPRs that commonly occur on gut enteroendocrine cells, adipocytes, and immune cells. Interventional studies presented that activation of GPR41 on enteroendocrine cells stimulated secretion of the gut hormone PYY, which functions to induce satiety and reduce food consumption [46], while by promoting GLP-1 secretion GPR43 signaling was proposed to mediate host insulin sensitivity [47]. Such molecular mechanisms of

SCFA-receptor-mediated metabolic responses were further established in human studies by the outcome that obese individuals who have administrated propionate increased the secretion of PYY and GLP-1 with significantly reduced adiposity and overall weight gain [48].

In support of the hypotheses outlined above, Jovanovski et al. aimed to summarize and quantify the effects of viscose fiber on body weight, body mass index (BMI), waist circumference, and body fat independent of calorie restriction, through a systematic review and meta-analysis of randomized controlled trials. Their results indicated that with an ad libitum diet, viscous fiber reduce the mean body weight, BMI and waist circumference while there was no change in body fat. Especially greater reductions in body weight was revealed in overweight and those with diabetes and metabolic syndrome. As they concluded, they stated dietary viscous fiber significantly improved body weight and other adiposity parameters independently of calorie restriction [49]. Another meta-analysis by Miketinas et al. supported the previous meta-analysis and they indicated that their primary aim was to assess the role of dietary fiber as a predictor of weight loss in participants who consumed calorie restricted diet for 6 months (−750 kcal/d from estimated energy requirement). Their results pointed out that dietary fiber intake, independently of macronutrient and energy intake, endorses weight loss and dietary adherence in adults with overweight or obesity consuming a calorie-restricted diet [50]. Salleh et al. carried out a systematic review which examined the effects of soluble dietary fiber using randomized controlled trials. As their study mentioned consumption of soluble fiber is advised since they slows gastric emptying, increases perceived satiety and acting a noteworthy role in appetite regulation. In their study, randomized controlled clinical studies conducted with different types of soluble fiber were examined in order to determine which type of fiber is more effective on weight loss. Their results indicated that polydextrose as a prebiotic dietary fiber candidate showed a significant reduction in energy intake yet compared to other types it was consumed in a higher doses (25 g) which prepared in liquid meal. This study shows that not all soluble fibers produce the same effect. They emphasized that further interventional studies are needed to determine whether combinations of these soluble fibers will have greater effects than individual fibers [51]. In another meta-analysis, specific species of prebiotics were evaluated and the efficiency of FOS and GOS prebiotics on body weight, BMI and fat mass were examined. According to the results, subjects consuming prebiotics demonstrated a significant reduction in body weight [52]. Another systematic review of randomized controlled trials indicated that, 5 (42%) of the 12 randomized controlled trial studies provided to the subjects nonviscous but fermentable fiber supplements in the form of manno-oligosaccharides, GOSs, and FOSs. According to the study results, soluble fiber supplementation reduced BMI, body weight, body fat compared with the effects of placebo treatments. Their study results concluded that soluble fiber supplementation improves both anthropometric and metabolic outcomes in overweight and obese adults [53]. Also the meta-analysis of randomized controlled trials indicated that viscous fiber within a calorie-restricted diet decreased body weight and body fat [54]. Overall, prebiotic dietary fibers modulate gut microbiota by increasing SCFAs and show effects in relation with adipose tissue, liver, brain, and pancreas. Thus, SCFAs stimulate satiety hormones such as GLP-1 and PYY, and shift glucose and lipid metabolism [55].

5. Conclusion

While all prebiotics are accepted as dietary fibers but not all dietary fibers are accepted as prebiotics. FOSs and GOSs are well known prebiotic dietary fiber since

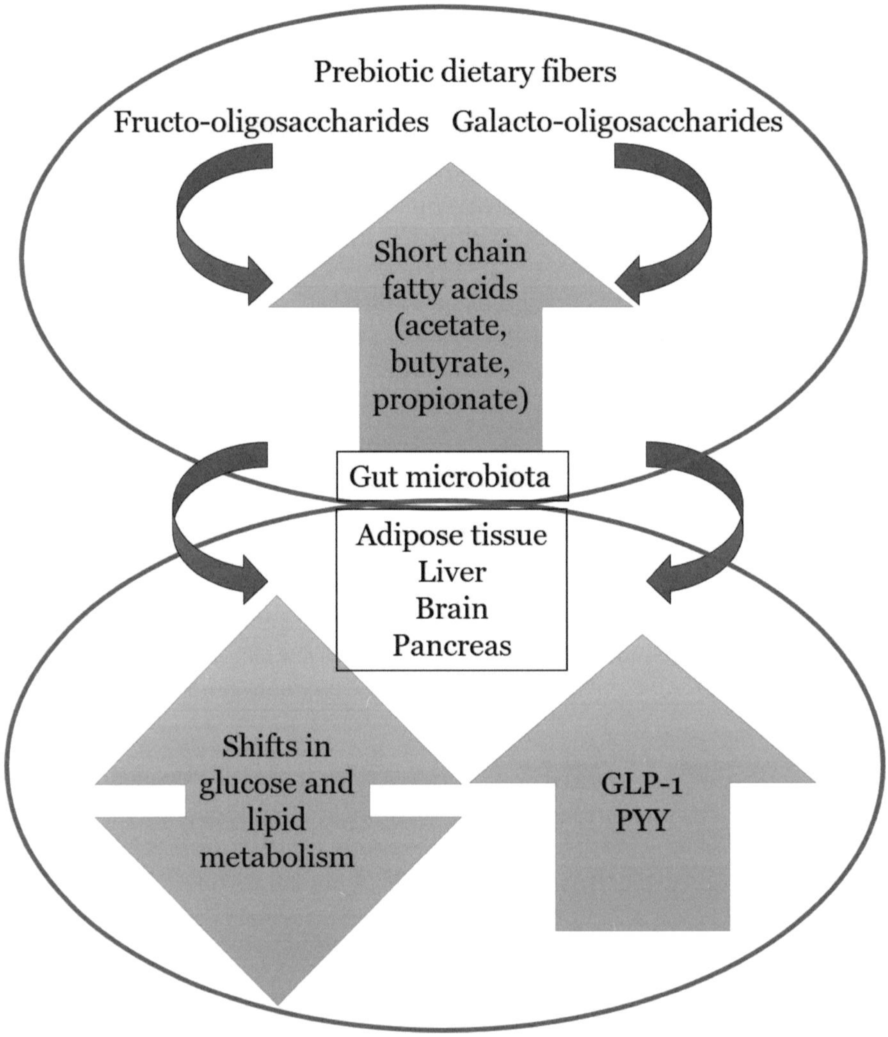

Figure 1.
Basic action of mechanisms of prebiotic dietary fibers on weight management.

studies mostly focus on them. These prebiotic dietary fibers has shown effect on weight management by increasing SCFAs thus modulating gut microbiota. SCFAs while modulating gut microbiota also stimulate satiety hormones such as GLP-1 and PYY, and shift glucose and lipid metabolism (**Figure 1**). Nevertheless, there is a need to further clinical trials to explain more comprehensively the effects of prebiotic dietary fibers on weight management.

Conflict of interest

The authors declare no conflict of interest.

Author details

Ceren Gezer* and Gözde Okburan
Department of Nutrition and Dietetics, Faculty of Health Sciences, Eastern Mediterranean University, Famagusta, North Cyprus via Mersin 10, Turkey

*Address all correspondence to: ceren.gezer@emu.edu.tr

References

[1] Kobyliak N, Conte C, Cammarota G, Haley A, Styriak I, Gaspar L et al. Probiotics in prevention and treatment of obesity: a critical view. Nutrition & Metabolism. 2016;13(1). DOI: 10.1186/s12986-016-0067-0

[2] Cerdó T, García-Santos J, G. Bermúdez M, Campoy C. The Role of Probiotics and Prebiotics in the Prevention and Treatment of Obesity. Nutrients. 2019;11(3):635. DOI:10.3390/nu11030635

[3] Li Y, Komarek A. Dietary fibre basics: Health, nutrition, analysis, and applications. Food Quality and Safety. 2017;1(1):47-59. DOI:10.1093/fqs/fyx007

[4] Meyer D. Health Benefits of Prebiotic Fibers. Advances in Food and Nutrition Research. 2015;74:47-91. DOI: 10.1016/bs.afnr.2014.11.002

[5] Verspreet J, Damen B, Broekaert W, Verbeke K, Delcour J, Courtin C. A Critical Look at Prebiotics Within the Dietary Fiber Concept. Annual Review of Food Science and Technology. 2016;7(1):167-190. DOI: 10.1146/annurev-food-081315-032749.

[6] Gibson G, Hutkins R, Sanders M, Prescott S, Reimer R, Salminen S et al. Expert consensus document: The International Scientific Association for Probiotics and Prebiotics (ISAPP) consensus statement on the definition and scope of prebiotics. Nature Reviews Gastroenterology & Hepatology. 2017;14(8):491-502. DOI: 10.1038/nrgastro.2017.75

[7] Slavin J. Fiber and Prebiotics: Mechanisms and Health Benefits. Nutrients. 2013;5(4):1417-1435. DOI: 10.3390/nu5041417

[8] Carlson J, Erickson J, Lloyd B, Slavin J. Health Effects and Sources of Prebiotic Dietary Fiber. Current Developments in Nutrition. 2018;2(3):1-8. DOI: 10.1093/cdn/nzy005

[9] Davani-Davari D, Negahdaripour M, Karimzadeh I, Seifan M, Mohkam M, Masoumi S et al. Prebiotics: Definition, Types, Sources, Mechanisms, and Clinical Applications. Foods. 2019;8(3):92. DOI:10.3390/foods8030092

[10] Rezende E, Lima G, Naves M. Dietary fibers as beneficial microbiota modulators: A proposal classification by prebiotic categories. Nutrition. 2021;89:111217. DOI: 10.1016/j.nut.2021.111217

[11] Stick R, Williams S. Disaccharides, Oligosaccharides and Polysaccharides. Carbohydrates: The Essential Molecules of Life. 2nd ed. Oxford: Elsevier; 2009. 321-341 p. DOI: 10.1016/B978-0-240-52118-3.00009-0

[12] Wilson B, Whelan K. Prebiotic inulin-type fructans and galacto-oligosaccharides: definition, specificity, function, and application in gastrointestinal disorders. Journal of Gastroenterology and Hepatology. 2017;32:64-68. DOI: 10.1111/jgh.13700

[13] Singh S, Jadaun J, Narnoliya L, Pandey A. Prebiotic Oligosaccharides: Special Focus on Fructooligosaccharides, Its Biosynthesis and Bioactivity. Applied Biochemistry and Biotechnology. 2017;183(2):613-635. DOI: 10.1007/s12010-017-2605-2

[14] Holscher H. Dietary fiber and prebiotics and the gastrointestinal microbiota. Gut Microbes. 2017;8(2):172-184. DOI: 10.1080/19490976.2017.1290756

[15] Liu F, Li P, Chen M, Luo Y, Prabhakar M, Zheng H et al. Fructooligosaccharide (FOS) and

Galactooligosaccharide (GOS) Increase Bifidobacterium but Reduce Butyrate Producing Bacteria with Adverse Glycemic Metabolism in healthy young population. Scientific Reports. 2017;7(1):11789. DOI: 10.1038/s41598-017-10722-2.

[16] Sudha ML, Rajeswari G, Venkateswara-Rao O. Effect of wheat and oat brans on the dough rheological and quality characteristics of instant vermicelli. Journal of Texture Studies. 2017;43: 195-202. DOI: 10.1111/j.1745-4603.2011.00329.

[17] Barber TM, Kabisch S, Pfeiffer AFH, Weickert MO. The health benefits of dietary fibre. Nutrients. 2020;12:10. DOI: 10.3390/nu12103209.

[18] Dahl WJ and Stewart ML. Position of the Academy of Nutrition and Dietetics: health implications of dietary fiber. Journal of the Academy of Nutrition and Dietetics. 2015;115: 1861-1870. DOI: 10.1016/j.jand.2015.09.003.

[19] Zhang A, Xu G, Liu D, Zhu W, Fan X, Liu X. Dietary fiber consumption and risk of stroke. European Journal of Epidemiology. 2013;28: 119-130. DOI: 10.1007/s10654-013-9783-1

[20] Dahm CC, Keogh RH, Spencer EA, Greenwood DC, Key TJ, Fentiman I et al. Dietary fiber and colorectal cancer risk: a nested case-control study using food diaries. Journal of National Cancer Institute. 2010;102: 614-626. DOI: 10.1093/jnci/djq092.

[21] World Cancer Research Fund/ American Institute for Cancer. (2011). Continuous update project report: food, nutrition, and physical activity and the prevention of colorectal cancer. http://www.dietandcancerreport.org/cancer_resource_center/downloads/cu/Colorectal-Cancer-2011-Report.pdf.

[22] Jiménez JP, Serrano J, Tabernero M, Arranz S, Rubio MED, Diz LG et al. Effects of group of dietary fiber in cardiovascular disease risk factors. Journal of Nutrition. 2008;24: 646-653. 10.1016/j.nut.2008.03.012.

[23] Wannamethee SG, Whincup PH, Thomas MC, Sattar N. Associations between dietary fiber and inflammation, hepatic function, and risk of type 2 diabetes in older men: potential mechanisms for the benefits of fiber on diabetes risk. Diabetes Care. 2008;32: 1823-1825. DOI: 10.2337/dc09-0477.

[24] Ye EQ, Chacko SA, Chou EL, Kugizaki M, Liu S. Greater whole-grain intake is associated with lower risk of type 2 diabetes, cardiovascular disease, and weight gain. Journal of Nutrition. 2012;142: 1304-1313. DOI: 10.3945/jn.111.155325.

[25] Cho SS, Qi L, Fahey GC, Klurfeld DM. Consumption of cereal fiber, mixtures of whole grains and bran, and whole grains and risk reduction in type 2 diabetes, obesity, and cardiovascular disease. American Journal of Clinical Nutrition. 2013;98: 594-619. DOI: 10.3945/ajcn.113.067629.

[26] Yao B, Fang H, Xu W. Dietary fiber intake and risk of type 2 diabetes: a dose response analysis of prospective studies. European Journal of Epidemiology. 2014;29: 79-88. DOI: 10.1007/s10654-013-9876

[27] McRae MP. Dietary fiber is beneficial for the prevention of cardiovascular disease: an umbrella review of meta-analyses. Journal of Chiropractic Medicine. 2016;16:4. DOI: 10.1016/j.jcm.2017.05.005.

[28] Fairbanks A, Blau K, Jorgensen MJ. High-fiber diet promotes weight loss and affects maternal behavior in Vervet monkeys. Journal of Primatology. 2010;72: 234-242. DOI: 10.1002/ajp.20772

[29] Mozaffarian D, Hao T, Rimm EB, Willett WC, Hu FB. Changes in diet and lifestyle and long-term weight gain in women and men. The New England Journal of Medicine. 2011;364: 2392-2404. DOI: 10.1056/NEJMoa10 14296.

[30] Wanders AJ, Borne JJGC, Graaf C, Hulshof T, Jonathan MC, Kristensen M, et al. Effects of dietary fiber on subjective appetite, energy intake and body weight: a systematic review of randomized controlled trials. Obesity Reviews. 2011;12: 724-739. DOI: 10.1111/j.1467-789X.2011.00895.

[31] Shay CM, van Horn L, Stamler J. Food and nutrient intakes and their associations with lower BMI in middle-aged US adults: the international study of macro-/Micronutrients and blood pressure (INTERMAP). American Journal of Clinical Nutrition. 2012;96: 483-491. DOI: 10.3945/ajcn.111. 025056

[32] Clark MJ, Slavin JL. The effect of fiber on satiety and food intake: a systematic review. Journal of the American College of Nutrition. 2013;32: 200-211. DOI: 10.1080/07315724. 2013.791194.

[33] Li SS, Kendall CWC, Souza RJ, Jayalath VH, Cozma AI, Ha V, et al. Dietary pulses, satiety and food intake: a systematic review and meta-analysis of acute feeding trials. Obesity. 2014;22: 1773-1780. DOI: 10.1002/ oby.20782.

[34] Watzl B, Girrbach S, Roller M. Inulin, oligofructose and immunomodulation. British Journal of Nutrition. 2005;93:49-55. 10.1079/ bjn20041357.

[35] Simpson HL, Campbell BJ. Review article: dietary fiber-microbiota interactions. Alimentary Pharmacology & Therapeutics. 2015;42(2): 158-179. DOI: 10.1111/apt.13248.

[36] Dong H, Sargent, LJ, Chatzidiakou Y, Saunders C, Harkness L, Bordenave N, et al. Orange pomace fiber increases a composite scoring of subjective ratings of hunger and fullness in healthy adults. Appetite. 2016;107: 478-485. 10.1016/j. appet.2016.08.118.

[37] Cronin P, Joyce SA, Toole PWO, Connor EMO. Dietary fibre modulates the gut microbiota. Nutrients. 2021;13: 1655. DOI: 10.3390/nu13051655.

[38] Fernandes R, Rosario VA, Mocellin MC, Kuntz MGF, Trindade EBSM. Effects of inulin-type fructans, galacto-oligosaccharides and related synbiotics on inflammatory markers in adult patients with overweight or obesity: a systematic review. Clinical Nutrition. 2017;36:5. DOI: 10.1016/j.clnu.2016.10.003.

[39] Khan MT, Nieuwdorp M, Backhed F. Microbial modulation of insulin sensitivity. Cell Metabolism. 2014: 4;20:5. DOI: 10.1016/j.cmet.2014.07.006.

[40] Delzenne NM, Olivares, M, Neyrinck AM, Beaumont M, Kjølbæk L, Larsen TM et al. Nutritional interest of dietary fiber and prebiotics in obesity: lessons from the MyNewGut consortium. Clinical Nutrition. 2020;39:414-424. DOI: 0.1016/j. clnu.2019.03.002

[41] Delzenne DM, Neyrinck AM, Cani PD. Modulation of the gut microbiota by nutrients with prebiotic properties: consequences for host health in the context of obesity and metabolic syndrome. Microbial Cell Factories. 2011;10(Suppl1):S10. DOI: 10.1186/1475-2859-10-S1-S10.

[42] Vallianou N, Stratigou T, Christodoulatos GS, Dalamaga M. Understanding the role of the gut microbiome and microbial metabolites in obesity and obesity-associated

metabolic disorders: current evidence and perspectives. Current Obesity Report. 2019;8:317-332. DOI: 10.1007/s13679-019-00352-2.

[43] Benton D and Young HA. Reducing calorie intake may not help you lose body weight. Perspectives on Psychological Science. 2017;12: 703-714. DOI: 10.1177/1745691617690878.

[44] Lim JJ, Poppitt SD. How satiating are the 'satiety' peptides: a problem of pharmacology versus physiology in the development of novel foods for regulation of food intake. Nutrients. 2019;11(7):1517. DOI: 10.3390/nu11071517.

[45] Morrison D, Preston T. Formation of short chain fatty acids by the gut microbiota and their impact on human metabolism. Gut Microbes. 2016;7(3):189-200. DOI: 10.1080/19490976.2015.1134082.

[46] Samuel B, Shaito A, Motoike T, Rey F, Backhed F, Manchester J et al. Effects of the gut microbiota on host adiposity are modulated by the short-chain fatty-acid binding G protein-coupled receptor, Gpr41. Proceedings of the National Academy of Sciences. 2008;105(43):16767-16772. DOI: 10.1073/pnas.0808567105.

[47] Tolhurst G, Heffron H, Lam YS, Parker HE, Habib AM, Diakogiannaki E. et al. Short-chain fatty acids stimulate glucagon-like peptide-1 secretion via the G-protein-coupled receptor FFAR2. Diabetes. 2012;61:364-371. DOI: 10.2337/db11-1019.

[48] Chambers ES, Viardot A, Psichas A, Morrison DJ, Murphy KG, Zac-Varghese SEK. Effects of targeted delivery of propionate to the human colon on appetite regulation, body weight maintenance and adiposity in overweight adults. Gut. 2015;64:1744-1754. DOI: 10.1136/gutjnl-2014-307913.

[49] Jovanovski E, Mazhar N, Komishon A, Khayyat R, Li D, Mejia SB. Short-chain fatty acids stimulate glucagon-like peptide-1 secretion via the G-protein-coupled receptor FFAR2. Diabetes. 2012;61(2):364-371. DOI: 10.2337/db11-1019.

[50] Miketinas DC, Bray GA, Beyl RA, Ryan DH, Sacks FM, Champagne CM. Fiber intake predicts weight loss and dietary adherence in adults consuming calorie-restricted diets: The POUNDS Lost (Preventing Overweight Using Novel Dietary Strategies) Study. Journal of Nutrition. 2019;149:1742-1748. DOI: 10.1093/jn/nxz117.

[51] Salleh SN, Fairus AAH, Zahary MN, Raj NB, Jalil AM. Unravelling the effects of soluble dietary fibre supplementation on energy intake and perceived satiety in healthy adults: evidence from systematic review and meta-analysis of randomised-controlled trials. Foods. 2019;8:15. DOI:10.3390.

[52] John GK, Wang L, Nanavati J, Twose C, Singh R, Mullin G. Dietary alteration of the gut microbiome and its impact on weight and fat mass: a systematic review and meta-analysis. Genes. 2018;9:167. DOI:10.3390.

[53] Thompson S, Hannon B, An R, Holscher H. Effects of isolated soluble fiber supplementation on body weight, glycemia, and insulinemia in adults with overweight and obesity: a systematic review and meta-analysis of randomized controlled trials. American Journal of Clinical Nutrition. 2017;106(6):1514-1528. DOI: 10.3945/ajcn.117.163246

[54] Jovanovski E, Mazhar N, Komishon A, Khayyat R, Li D, Blanco Mejia S et al. Effect of viscous fiber supplementation on obesity indicators in individuals consuming calorie-restricted diets: a systematic review and meta-analysis of randomized controlled

trials. European Journal of Nutrition. 2020;60(1):101-112. DOI: 10.1007/s00394-020-02224-1.

[55] Delgado GTC, Tamashirob WMSC. Role of prebiotics in regulation of microbiota and prevention of obesity. Food Research International. 2018;113:183-188. DOI:10.1016/j.foodres.2018.07.013.

Chapter 6

Signaling Pathways Associated with Metabolites of Dietary Fibers Link to Host Health

Kavita Rani, Jitendra Kumar, Sonia Sangwan, Nampher Masharing, Murli Dhar Mitra and H.B. Singh

Abstract

Food is a basic requirement for human life and well-being. On the other hand, diet is necessary for growth, health and defense, as well as regulating and assisting the symbiotic gut microbial communities that inhabit in the digestive tract, referred to as the gut microbiota. Diet influences the composition of the gut microbiota. The quality and quantity of diet affects their metabolism which creates a link between diet. The microorganisms in response to the type and amount of dietary intake. Dietary fibers, which includes non-digestible carbohydrates (NDCs) are neither neither-digested nor absorbed and are subjected to bacterial fermentation in the gastrointestinal tract resulting in the formation of different metabolites called SCFAs. The SCFAs have been reported to effect metabolic activities at the molecu-larlevel. Acetate affects the metabolic pathway through the G-protein-coupled receptor (GPCR) and free fatty acid receptor2 (FFAR2/GPR43) while butyrate and propionate transactivate the peroxisome proliferator-activated receptors (PPARγ/NR1C3) and regulate the PPARγ target gene *Angptl4* in colonic cells of the gut. The NDCs via gut microbiota dependent pathway regulate glucose homeostasis, gut integrity and hormone by GPCR, NF-kB, and AMPK-dependent processes. In this chapter, we will focus on dietary fibers, which interact directly with gut microbes and lead to the production of metabolites and discuss how dietary fiber impacts gut microbiota ecology, host physiology, and health and molecule mechanism of dietary fiber on signaling pathway that linked to the host health.

Keywords: dietary fibe, gut microbiota ecology, host health, signaling pathway, molecule mechanism

1. Introduction

The human gut harbors a plethora of a complex community of micro-organisms that are vital for host development and physiology. This community of microbes inhabiting the gut called "gut microbiota" represents a mutualistic symbiotic relationship with the host [1]. The host creates a stable environment for the microbes while the microbes offer the host with an array of functions such as digestion of complex dietary macronutrients, minerals and vitamins production, pathogen protection, and immune system maintenance. Studies have shown that the gut microbiota comprises of about 3.8×10^{13} microorganisms [2] belonging to a wide spectrum of about 160 recognized gut bacterial species [3]. Generally, the opus of the gut microbiota is observed to be comparable in all healthy individuals, however the presence of different microbial species is determined by an individual's dietary habits, dietary patterns and lifestyle [4]. Dietary fibers (DFs) are vital modulators of the gut microbiota composition which directly impacts individual biological processes and homeostasis via the metabolites, a consequent of microbial fermentation of nutrients such as, short-chain unsaturated fats (SCFAs) [5]. The gut microbiota plays a key and essential role in the metabolization of DFs including non-digestible carbohydrates (NDCs), proteins and peptides, which has escape digestion by host enzymes in the upper gut and absorption in the lower digestive tract. These dietary constituents, are then subjected to fermentation by the microbiota in the cecum and colon (Macfarlane and Macfarlane, 2012) resulting in the production different metabolites called SCFAs varying in carbon number which includes mainly acetate (60%), propionate (25%) butyrate (15%) and methane (CH_4), carbon dioxide (CO_2) gases [6] which are known to have beneficial effects by behaving as signaling molecules via different pathways. From among the different SCFAs produced Acetate is the most abundant and it is used by many gut commensals to produce propionate and butyrate in a growth-promoting cross-feeding process. Moreover, the SCFA, have been shown a to regulate metabolic activities. Acetate affects the metabolic pathway through the G protein-coupled receptor (GPCR) and free fatty acid receptor 2 (FFAR2/GPR43) while butyrate and propionate transactivate the peroxisome proliferator-activated receptorsγ (PPARγ/NR1C3) and regulates the PPARγ target gene *Angptl4* in colonic cells of the gut. The FFAR2 signaling pathway regulates the insulin-stimulated lipid accumulation in adipocytes and inflammation however peptide tyrosine-tyrosine and glucagon-like peptide 1 regulate appetite. The NDCs via microbiota dependent pathway regulate glucose homeostasis, gut integrity, and hormone by GPCR, NF-kB, and AMPK-dependent processes. Hence in this chapter, emphasis is given to address the effects of dietary fibers metabolites as prime signaling molecules, through different signaling pathways and their link between gut microbiota and the host health.

2. Dietary fibers (DFs), gut microbiota and SCFAs metabolites

2.1 Dietary fibers (DFs)

Dietary fibles defined by codex alimentarius commission (2009) are edible carbohydrate polymers with varying monomeric units that are impervious to the host digestive enzymes and thus has escape absorption in the small intestines. These includes, (1) edible naturally occurring carbohydrate polymers present in foods such as fruits, vegetables, legumes, and cereals (2) edible carbohydrate polymers obtained from food raw materials by physical, enzymatic, and chemical means and (3) synthetic carbohydrate polymers. In addition, DFs are further divided either into polysaccharides (non-starchpolysaccharides [NSPs], resistant starch [RS], and resistant oligosaccharides [ROs]) or into insoluble and soluble forms [7]. Soluble fibers are fermented by the gut bacteria giving rise to metabolites such as short-chain fatty acids (SCFAs), insoluble forms of fibers such as cellulose and hemi-cellulose may or slowly digested by the gut bacteria and contributes to a fecal bulking effect, as they reach the colon. Delay absorption of glucose and lipids influencing post-prandial metabolism on the other hand are caused by most soluble NSPs, especially polymers with high molecular weight such as guar gum, certain pectins, b-glucans, and psyllium, are viscous, meaning that they are able to form a gel structure in the intestinal tract that can [7]. Food sources such as legumes, vegetables, nuts, seeds, fruits, and cereals forms the sources of soluble and insoluble fibers whereas RS can only be found in starchy foods such as legumes, tubers, cereals, and fruitlike green bananas, whereas pectin's are more abundant in fruits and some vegetables, whereas b-glucans are found in cereals [8]. Recently, due to low consumption of DFs in the Industrialized Western world Fortification of foods with extracted or synthesized non-digestible carbohydrates is carried out as a strategy to increase fiber intake. Besides, a wide range of commercial DFs are currently available [9] worldwide, called "prebiotics" on the fact that they exert health benefits by selectively inducing beneficial bacterial populations in the gut. However, contrastingly, studies have reported that irrespective of the types of fibers, virtually all fibers will induce specific shifts in microbiota composition as a result of competitive interactions, and which of these compositional shifts may be beneficial for health, has not yet been established [10]. Furthermore, the mechanisms that have been established to be beneficial to health, is not calculative on the selective utilization of the carbohydrates but on an integrative effect of bacterial fermentation, producing metabolic compounds (e.g., SCFAs) [11], physiological changes (pH lowering), or protection of the mucus layer [12, 13]. Hence, a change in the emphasis of the prebiotic concept away from the selective effect of specific dietary components on gut microbial communities towards the effects of ecological and functional consequences of fiber fermentation, is more significant for host physiology and health [10].

2.2 Gut microbiota

Microorganisms including several species of bacteria, yeast, and viruses make up the Gut microbiota. Out of the different Bacterial phyla, a few phyla represented, by about 160 species [14] composed the gut microbiota. Some of the dominant gut microbial phyla are *Firmicutes, Bacteroidetes, Actinobacteria, Proteobacteria, Fusobacteria, and Verrucomicrobia,* with phyla Firmicutes and phyla Bacteroidetes [15] making up to 90% of gut microbiota. *Clostridium, Enterococcus, Lactobacillus, Bacillus, and Ruminicoccus* are among the more than 200 genera in the Firmicutes phylum. *Clostridium* genera represent 95% of the Firmicutes phyla. Phylum Bacteroidetes consists of Bacteroides and Prevotella as predominant genera. The Actinobacteria a less abundant phylumis mainly represented by the Bifidobacterium genus [15]. Besides, the gut microbiome is to a very large extent affected by dietary administration of fiber, which alters the gut microbiota by providing substrates for microbial growth, and expansion of their populations [7]. The possession of different enzymes, about 130 glycoside hydrolase, 22 polysaccharidelyase, and 16 carbohydrate esterase enzyme families, allows the gut microbiome to switch between different energy sources of fibers depending on their availability [16]. Bacteria such as Firmicutes and Actinobacteria has been found to be prime species, which initiates the degradation of complex substrates [7]. Species such as *Bifidobacteriumadolescentis*, *Ruminoccocusbromii*, *Eubacteriumrectale*, and *Parabacteroides distasonis*play significant roles in degrading resistant Starch [1, 17]. The consumption of galactooligosaccharides mainly induces Bifidobacterium species possessing the enzymatic machinery to utilize the substrate [18]. Reports have also suggested that, degradation of complex substates, occurs in a cascade where, different species will contribute equally at different stages towards production of metabolites [7]. Primary fiber degraders are species that initiate the utilization of a complex fiber through what can be considered a "guild" of species [19] or a keystone species. Although *R. bromii* does not make butyrate, it is considered a keystone species for the breakdown of RS and contributes significantly to butyrate generation in the colon. Other dietary fiber types are expected to have similar keystone species, although they have yet to be discovered.

2.3 SCFAs metabolites

Dietary fibers, are metabolized by the microbiota in the cecum and colon [20] resulting in the formation of major products such as particular, acetate, propionate, and butyrate [21]. However, studies have reported that, microbes can utilize amino acids from dietary proteins and triglycerols from fats [22, 23] to facilitate diminished supply of dietary fermentable fibers resulting in reduced fermentative activity and formation of SCFAs as minor end products [24]. Although, protein fermentation was observed to the SCFA pool but, however dietary proteins mostly give rise to branched-chain fatty acids such asisobutyrate, 2-methylbutyrate, and isovalerate, [25] which are may have a concerning effect as a result of insulin resistance [26].

Acetate (C2) is a major SCFA metabolite produced from pyruvate. Many gut bacteria produce Acetate from pyruvate via acetyl-CoA or the Wood-Ljungdahl pathway, which produces acetate via two branches: (1) the C1-body branch (also known as the Eastern branch) via CO_2 reduction to formate and (2) the carbon monoxide branch (also known as the Western branch) via CO_2 reduction to CO, which is then combined with a methyl group. Propionate is created when succinate is converted to methylmalonyl-CoA through the succinate pathway. Furthermore, propionate, can also be synthesized from acrylate using lactate as a precursor via the acrylate pathway [27] and via the propanediol pathway using deoxyhexose sugars as substrates [28]. Butyrate, the third main SCFA, is produced by the condensation of two molecules of acetyl-CoA and subsequent reduction to butyryl-CoA, which can then be converted to butyrate by phosphotrans butyrylase and butyrate kinase via the classical pathway [29]. The butyryl-CoA: acetate CoA-transferase enzyme can also convert butyryl-CoA to butyrate [30]. Besides, reports have also shown that some microbes can use both lactate and acetate to synthesize butyrate. Butyrate can also be produced from proteins via the lysine pathway, according to a recent analysis of metagenome data [31], implying that microorganisms in the gut can adjust to dietary changes in order to sustain the synthesis of important metabolites like SCFAs. SCFA levels vary along the length of the gut, with the highest concentrations in the cecum and proximal colon and decreasing towards the distal colon [21].

3. Dietary fibers metabolites signaling mechanism and their health implications

3.1 Molecular mechanism of dietary fibers (DFs) and its metabolites

The metabolites of dietary fibers (DFs) are SCFAs that play a significant role in metabolic diseases prevention and treatment along with some contradictory research finding [32]. The SCFAs formate, lactate, acetate, propionate, and butyrate are produces by the saccharolytic fermentation of the dietary fibrous [33] which have a significant role in the maintenance of health by reducing the chances of development of different disease.

World Health Organization have recommended daily intake of dietary fiber 20 g per 1000 kcal consumed for adults human being and this (20 g per 1000 kcal) quantity of dietary fiber is full filled by the daily consumption 400 g per day of fresh vegetables, fruits and grains (https://www.who.int/news-room/fact-sheets/detail/healthy-diet). Modern life style, dietary pattern, seasonality, stress, habitat, consumption of antibiotics and disease cause a drastic change in dietary pattern of individual's finally leads to gut microbial alteration [34] that influence production of SCAFs. The various physiological functions in the gut (including adding the energy to colonocytes, maintaining their mobility, blood flow, and regularize the movements of electrolytes and nutrients within the lumen) activate and modulate by SCAFs [35]. Colonic cell proliferation, differentiation and integrity mentioned by butyrate along with the major and preferred metabolic substrate for colonocytes 60–70% energy requirement [36]. Propionate maintains glucose homeostasis by gluconeogenic pathway [37]. The expression of leptin has enhanced by propionate and acetate. Leptin is a potent anorectic hormone, in adipocytes [38]. Acetate is a lipogenic SCFAs, reduced levels of acetate would result in decreased lipogenesis [37]. In the rat hepatocytes, acetate act as de novo lipogenesis and cholesterol synthesis, and these two pathways are to be inhibited by propionate [39]. The increased levels of propionate SCFAs would assist in the inhibition of acetate conversion into lipid in adipose tissue and the liver. The DFs via gut microbiota increase the rate of acetate synthesis while reducing the level of propionate in cells [40]. Acetate SCFAs is inversely related to plasma insulin levels [41] and acetate also activates leptin secretion in murine adipocytes [42].

High-fat diet-fed rats have increased acetate (C2) production due to gut microbiota that leads to ghrelin secretion and glucose-stimulated insulin secretion by activation of the parasympathetic nervous system (PNS), apart from these high calorically dense diet through gut microbiota-brain-β-cell axis promotes obesity and health complications by regimenting glucose and lipids homeostasis [43].

New study finding by many researchers groups have suggested that [44, 45] the loss of gut microbiota species from the colonic microbiota is associated to consumption of the high-fat, low-dietary fiber diets and other nutrient intake and diversity of gut microbiome [46, 47]. The fermentable dietary fibers directly govern the diversity of the gut microbiota [48], SCFAs regulate the different physiological activity of host. The majority of SCFAs transported across the mucosa by active transport, mediated by two receptors. The monocarboxylate transporter 1 (MCT-1) and the sodium-coupled monocarboxylate transporter 1 (SMCT-1) receptors. Direct inhibition of histone deacetylases HDACs to directly regulate gene expression and SCFA also effects signaling through G-protein-coupled receptors (GPCRs), this may influence host physiology by modulate biological responses of the host.

3.2 SCFAs sensing signaling pathway

All physiological activities occurring in the body are gut metabolites driven and SCFAs are connecting the link between the gut immunity with microbiota. The crucial role of SCFA has been signified in shaping and regulating both local and peripheral immune systems that respond to host metabolism via inflammatory pathways. Therefore, SCFAs modulate functions of the different systems including the enteric, nervous, endocrine, and blood vascular system serving as a key factor to regulate metabolic disorders and immunity. The dietary fibers metabolites exerted effects via their receptors, like the G protein-coupled receptor (FFAR3/GPR41 and FFAR2/GPR43 and GPR109a) through the inhibition of histone deacetylases and the activation of G-protein coupled receptors [32, 49].

3.3 SCFAs sensing signaling pathway in immunological responses

Gut bacteria produced SCFAs from indigestible saccharides diet precursors and SCFAs transported across the mucosa by active transport mediated by two receptors, monocarboxylate transporter 1 (MCT-1) and sodium-coupled monocarboxylate transporter 1 (SMCT-1) receptors which influence host physiological functions and modulate biological responses of the host. The main mechanism is direct inhibition of histone deacetylases (HDACs) to directly regulate gene expression. HDACs remove acetyl groups (deacetylation) from lysine residues of histones [50]. Transcription of genes is enhanced through inhibition of HDACs function by increasing histone acetylation. Dietary fibers SCFAs inhibit HDACs activity and therefore suppress expression of gene in different cells. Butyrate (C4) SCFAs is the most potent inhibitor of HDACs activity and induces gene activation by facilitating the access of transcription factors to promoter region, such kind's activity of C4 known as an epigenetic modification of chromatins [51]. The SCFAs-mediated HDACs inhibition, acts as an anti-inflammatory immune response mediated by less production of inflammatory cytokines IL-8, IL-6, and TNFα [52]. Apart from these butyrate and propionate reduced NF-kB activity and inflammatory cytokines [53], showing that the anti-inflammatory effects of SCFAs are mediated through the modulation of NF-kB signaling pathway. Beside this the SCFAs also affect signaling through GPCRs. The SCFAs activate different GPCRs e.g. propionate (C3) is a most potent activator of GPR43. The expression of GPR43 has been reported in the entire gastrointestinal tract (GIT) along with cells of the immune and nervous systems. In GIT, GPR43 is highly expressed in endocrine L-cells of the ileum and colon of intestinal PYY and GLP-1 [54] producing cells as well as on colonocytes and enterocytes. The order of potency was reported as like propionate >butyrate>acetate for GPR41 receptor [55]. The SCFAs control the body weight through the release of leptin in adipose tissue by the expression of GPR41 [56]. The SCFAs play crucial role in metabolic functions of hepatic cells through the FFAR3 signaling pathway without influencing the intestinal environment [57]. Niacin receptor 1 (GPR109a) is activated by C4 at low concentration while highly expressed in adipocytes with a lesser extent is also expressed on immune cells. Activation of GPR109a in adipocytes suppresses lipolysis and the lowering of plasma-free fatty acid levels (FFAs) [58]. Through epigenetic mechanisms via histone acetylation acetate also increases fatty acid synthesis [59]. Therefore these finding could helpful to promote the development of functional foods using SCFAs or dietary significance of non-digestible carbohydrates fiber.

3.4 SCFAs sensing signaling pathway in hormone regulation

Gut microbes regulates the host metabolism by secretion of gut hormone. Gut microbiota induced signal to nearby intestinal enteroendocrine cells through microbial metabolites of DFs. These enteroendocrine cells release metabolically active hormones like GLP-1, PYY, GIP, 5-HT, and CCK which influence feeding behavior, glucose metabolism, insulin sensitivity and adiposity. Dietary components also impact on the composition of gut microbiota which may have further downstream consequences on gut hormone secretion and host metabolism. Enterochromaffin cells (EC) of the gut are the main source of serotonin (5-HT, 5-hydroxytryptamine). The EC is dispersed throughout the GI tract of the host and constitutes about half of all enteroendocrine cells. The gut microbiota influences 5-HT levels in the host. The antibiotic-treated mice study showed that significantly lower levels of EC cell-derived 5-HT when compared to antibiotic free animals. The EC cells can sense microbial metabolites by FFAR2 and FFAR3 signaling mechanisms [60]. PYY (Peptide tyrosine-tyrosine) regulates food intake and satiety through activation of central G protein-coupled Y2 receptors on neuropeptide Y (NPY) and AgRP neurons in arcuate nucleus of hypothalamic part of brain [61]. This is the relay of signaling cascade where by appetite-stimulating NPY neurons are suppressed that allowing for the activation of the satiety-inducing a-MSH / pro-opiomelanocortin (POMC) pathway [62]. The ability of gut microbiota to influence PYY secretion, therefore, gut microbiota has significant implications for the development of metabolic disease and obesity. Study reported that oral administration of C4 increased circulating PYY level [63] by FFAR2/3 signaling. Glucagon-like peptide 1 (GLP-1) augments insulin and inhibits glucagon secretion from the pancreas cells. GLP-1 inhibits gastric emptying and influences satiety and food intake [64, 65]. Orally administrated sodium butyrate in mice has been shown to transiently increase GIP and GLP-1 secretion and GIP level were associated with adiposity reported the ileal infusion of acetate, propionate, and butyrate during feeding in pigs, increased plasma CCK levels and paradoxically inhibit pancreatic secretion [66]. SCFAs are influence insulin function via their receptors [67, 68]. Glucose homeostasis in type 2 diabetes mellitus patients managed by fiber reaches diet that alters the gut microbiota. The deficiency in SCFAs production in host has associated with type 2 diabetes by interfering HbA1c levels in circulation [69]. Diet plat a major role in gut microbiota composition and gut microbiota regulates metabolism via metabolites produces by plant-based diets and intake of probiotics increases secretion of carbohydrate-active enzymes [70] in luminal of GIT.

4. Conclusion

Dietary fibers and its gut microbial metabolite SCFAs have been known to exert metabolic benefits to the host [71]. Various health benefits have been reported whereby Dietary fibers via SCFA increase plasma SCFA levels to active FFAR3 which has been shown to improve hepatic metabolic conditions. Furthermore, Dietary fibers consumption reduced HFD-induced liver weight growth and hepatic TG accumulation, as well as a shift in hepatic lipid metabolism. Dietary SCFAs consumption improved hepatic metabolic conditions via the FFAR3 signaling pathway. Besides, Dietary fibers were reported to shift the gut microbiome towards the production of more butanoate which is accompanied by up-regulation of microbiota and AMP-activated protein kinase (AMPK)-dependent gene expression which contributes to intestinal integrity and homeostasis by affecting metabolism, transporter expression. In addition, microbial metabolite SCFAs derived from microbial fermentation of dietary fibers are likely to have more broad impacts on various aspects of host physiology including health. Hence, Diet plays a pivotal role and is key as they have a significant impact on the composition, variety, and richness of the gut microbiota which directly determines the formation of essential SCFAs metabolites. Different aspects of the diet have a time-dependent effect on gut bacterial ecosystems. Long-term dietary patterns, particularly high protein and animal fat intake, have been demonstrated to diminish the number of beneficial microorganisms, which has been linked to host health.

Conflict of interest

The authors declare no competing interest.

Author contributions

The authors' responsibilities were as follows- Kavita Rani, Jitendra Kumar, S. Sangwan, Nampher Masharing, M.D. Mitra and H.B. Singh conceived and designed the chapter. Draft was completed by Kavita Rani, Jitendra Kumar, Nampher Masharing.

Notes/thanks/other declarations

We would like to acknowledge the support from the Director, ICAR-National Dairy Research Institute, Karnal-132,001, Haryana, India.

Author details

Kavita Rani[1*†], Jitendra Kumar[1†], Sonia Sangwan[1], Nampher Masharing[1], Murli Dhar Mitra[2] and H.B. Singh[3]

1 ICAR-National Dairy Research Institute, Karnal, Haryana, India

2 Department of Chemistry, Indian Institute of Technology (IIT), Banaras Hindu University (BHU), Varanasi, Uttar Pradesh, India

3 Department of Chemistry, Indian Institute of Technology (IIT), Hauz Khas South Delhi, Delhi, India

*Address all correspondence to: kavitamalik.ndri@gmail.com

† Both authors are equally contributed to this work.

References

[1] Walker AW, Ince J, Duncan SH, Webster LM, Holtrop G, Ze X, Brown D, Stares MD, Scott P, Bergerat A, Louis P. Dominant and diet-responsive groups of bacteria within the human colonic microbiota. The ISME journal. 2011; 5(2):220-230.

[2] Jandhyala SM, Talukdar R, Subramanyam C, Vuyyuru H, Sasikala M, Reddy DN. Role of the normal gut microbiota. World journal of gastroenterology: WJG. 2015 7;21(29):8787.

[3] Laterza L, Rizzatti G, Gaetani E, Chiusolo P, Gasbarrini A. The gut microbiota and immune system relationship in human graft-versus-host disease. Mediterranean journal of hematology and infectious diseases. 2016; 8(1).

[4] Sender R, Fuchs S, Milo R. Revised estimates for the number of human and bacteria cells in the body. PLoS biology. 2016 19;14(8):e1002533.

[5] Gentile CL, Weir TL. The gut microbiota at the intersection of diet and human health. Science. 2018 Nov 16;362(6416):776-780.

[6] Tazoe H, Otomo Y, Kaji I, Tanaka R, Karaki SI, Kuwahara A. Roles of short-chain fatty acids receptors, GPR41 and GPR43 on colonic functions. J Physiol Pharmacol. 2008 1;59 (Suppl 2):251-62.

[7] Deehan EC, Duar RM, Armet AM, Perez-Munoz ME, Jin M, Walter J. Modulation of the gastrointestinal microbiome with nondigestible fermentable carbohydrates to improve human health. Microbiology spectrum. 2017;5(5):5-.

[8] Lovegrove A, Edwards CH, De Noni I, Patel H, El SN, Grassby T, Zielke C, Ulmius M, Nilsson L, Butterworth PJ, Ellis PR. Role of polysaccharides in food, digestion, and health. Critical reviews in food science and nutrition. 2017;57(2):237-253.

[9] Deehan EC, Walter J. The fiber gap and the disappearing gut microbiome: implications for human nutrition. Trends in Endocrinology & Metabolism. 2016; 27(5):239-242.

[10] Bindels LB, Delzenne NM, Cani PD, Walter J. Towards a more comprehensive concept for prebiotics. Nature reviews Gastroenterology & hepatology. 2015; 12(5):303.

[11] Koh A, De Vadder F, Kovatcheva-Datchary P, Bäckhed F. From dietary fiber to host physiology: short-chain fatty acids as key bacterial metabolites. Cell. 2016; 165(6):1332-1345.

[12] Schroeder BO, Birchenough GM, Ståhlman M, Arike L, Johansson ME, Hansson GC, Bäckhed F. Bifidobacteria or fiber protects against diet-induced microbiota-mediated colonic mucus deterioration. Cell host & microbe. 2018; 23(1):27-40.

[13] Zou J, Chassaing B, Singh V, Pellizzon M, Ricci M, Fythe MD, Kumar MV, Gewirtz AT. Fiber-mediated nourishment of gut microbiota protects against diet-induced obesity by restoring IL-22-mediated colonic health. Cell host & microbe. 2018; 23(1): 41-53.

[14] Laterza L, Rizzatti G, Gaetani E, Chiusolo P, Gasbarrini A. The gut microbiota and immune system relationship in human graft-versus-host disease. Mediterranean journal of hematology and infectious diseases. 2016; 8(1).

[15] Arumugam M, Raes J, Pelletier E, Le Paslier D, Yamada T, Mende DR, Fernandes GR, Tap J, Bruls T, Batto JM, Bertalan M. Enterotypes of the human gut microbiome. nature. 2011;473 (7346):174-180.

[16] Flint HJ, Scott KP, Duncan SH, Louis P, Forano E. Microbial degradation of complex carbohydrates in the gut. Gut microbes. 2012;3(4): 289-306.

[17] Martínez I, Kim J, Duffy PR, Schlegel VL, Walter J. Resistant starches types 2 and 4 have differential effects on the composition of the fecal microbiota in human subjects. PloS one. 2010;5(11):e15046.

[18] Davis LM, Martínez I, Walter J, Goin C, Hutkins RW. Barcoded pyrosequencing reveals that consumption of galactooligosaccharides results in a highly specific bifidogenic response in humans. PLoS One. 2011;6(9):e25200.

[19] Zhao L, Zhang F, Ding X, Wu G, Lam YY, Wang X, Fu H, Xue X, Lu C, Ma J, Yu L. Gut bacteria selectively promoted by dietary fibers alleviate type 2 diabetes. Science. 2018 ; 359(6380):1151-1156.

[20] Macfarlane GT, Macfarlane S. Bacteria, colonic fermentation, and gastrointestinal health. Journal of AOAC International. 2012 Jan 1;95(1):50-60.

[21] Cummings J, Pomare EW, Branch WJ, Naylor CP, Macfarlane GT. Short chain fatty acids in human large intestine, portal, hepatic and venous blood. Gut. 1987 Oct 1;28(10):1221-1227.

[22] Cummings JH, Macfarlane GT. The control and consequences of bacterial fermentation in the human colon. Journal of Applied Bacteriology. 1991;70(6):443-459.

[23] Wall R, Ross RP, Shanahan F. O Mahony L, O Mahony C, Coakley M, et al. Metabolic activity of the enteric microbiota influences the fatty acid composition of murine and porcine liver and adipose tissues. Am J Clin Nutr. 2009;89(5):1393-1401.

[24] Russell WR, Gratz SW, Duncan SH, Holtrop G, Ince J, Scobbie L, Duncan G, Johnstone AM, Lobley GE, Wallace RJ, Duthie GG. High-protein, reduced-carbohydrate weight-loss diets promote metabolite profiles likely to be detrimental to colonic health. The American journal of clinical nutrition. 2011;93(5):1062-1072.

[25] Smith EA, Macfarlane GT. Dissimilatory amino acid metabolism in human colonic bacteria. Anaerobe. 1997;3(5):327-337.

[26] Newgard CB, An J, Bain JR, Muehlbauer MJ, Stevens RD, Lien LF, Haqq AM, Shah SH, Arlotto M, Slentz CA, Rochon J. A branched-chain amino acid-related metabolic signature that differentiates obese and lean humans and contributes to insulin resistance. Cell metabolism. 2009 ;9(4):311-326.

[27] Hetzel M, Brock M, Selmer T, Pierik AJ, Golding BT, Buckel W. Acryloyl-CoA reductase from Clostridium propionicum: An enzyme complex of propionyl-CoA dehydrogenase and electron-transferring flavoprotein. European journal of biochemistry. 2003;270(5):902-910.

[28] Scott KP, Martin JC, Campbell G, Mayer CD, Flint HJ. Whole-genome transcription profiling reveals genes up-regulated by growth on fucose in the human gut bacterium "Roseburia inulinivorans". Journal of bacteriology. 2006; 188(12):4340-4349.

[29] Louis P, Duncan SH, McCrae SI, Millar J, Jackson MS, Flint HJ. Restricted distribution of the butyrate kinase pathway among butyrate-producing bacteria from the human colon. Journal of bacteriology. 2004;186(7):2099-2106.

[30] Duncan SH, Barcenilla A, Stewart CS, Pryde SE, Flint HJ. Acetate utilization and butyryl coenzyme A (CoA): acetate-CoA transferase in butyrate-producing bacteria from the human large intestine. Applied and environmental microbiology. 2002;68(10):5186-5190.

[31] Vital M, Howe AC, Tiedje JM. Revealing the bacterial butyrate synthesis pathways by analyzing (meta) genomic data. MBio. 2014;5(2):e00889-e00814.

[32] Tirosh A, Calay ES, Tuncman G, Claiborn KC, Inouye KE, Eguchi K, Alcala M, Rathaus M, Hollander KS, Ron I, Livne R. The short-chain fatty acid propionate increases glucagon and FABP4 production, impairing insulin action in mice and humans. Science translational medicine. 2019;11(489).

[33] Macfarlane S, Macfarlane GT. Regulation of short-chain fatty acid production. Proceedings of the Nutrition Society. 2003;62(1):67-72.

[34] King DE, Mainous III AG, Lambourne CA. Trends in dietary fiber intake in the United States, 1999-2008. Journal of the Academy of Nutrition and Dietetics. 2012;112(5):642-648.

[35] Tazoe H, Otomo Y, Kaji I, Tanaka R, Karaki SI, Kuwahara A. Roles of short-chain fatty acids receptors, GPR41 and GPR43 on colonic functions. J Physiol Pharmacol. 2008;59(Suppl 2):251-262.

[36] Hong MY, Turner ND, Murphy ME, Carroll RJ, Chapkin RS, Lupton JR. In vivo regulation of colonic cell proliferation, differentiation, apoptosis, and P27Kip1 by dietary fish oil and butyrate in rats. Cancer prevention research. 2015;8(11):1076-1083.

[37] den Besten G, Lange K, Havinga R, van Dijk TH, Gerding A, van Eunen K, Muller M, Groen AK, Hooiveld GJ, € Bakker BM, et al. Gut-derived short-chain fatty acids are vividly assimilated into host carbohydrates and lipids. Am J Physiol Gastrointest Liver Physiol 2013; 305: G900-10; PMID: 24136789;http:// dx.doi.org/10.1152/ ajpgi.00265.2013

[38] Zaidi MS, Stocker CJ, O'Dowd J, Davies A, Bellahcene M, Cawthorne MA, Brown AJ, Smith DM, Arch JR. Roles of GPR41 and GPR43 in leptin secretory responses of murine adipocytes to short-chain fatty acids. FEBS letters. 2010;584(11):2381-2386.

[39] Demigne C, Morand C, Levrat MA, Besson C, Moundras C, Remesy C. Effect of propionate on fatty acid and cholesterol synthesis and on acetate metabolism in isolated rat hepatocytes. Br J Nutr 1995; 74:209-219; PMID: 7547838; http://dx.doi.org/10.1079/ BJN19950124.

[40] Liou AP, Paziuk M, Luevano JM, Jr., Machineni S, Turnbaugh PJ, Kaplan LM. Conserved shifts in the gut microbiota due to gastric bypass reduce host weight and adiposity. Sci Transl Med 2013; 5:178ra41; PMID: 23536013; http://dx.doi.org/10.1126/scitranslmed.3005687

[41] Layden BT, Yalamanchi SK, Wolever TM, Dunaif A, Lowe WL, Jr. Negative association of acetate with visceral adipose tissue and insulin levels. Diabetes Metab Syndr Obes 2012; 5:49-55; PMID:22419881; http://dx.doi.org/ 10.2147/DMSO.S29244

[42] Zaibi MS, Stocker CJ, O'Dowd J, Davies A, Bellahcene M, MA, Brown AJ, Smith DM, Arch JR. Roles of GPR41 and GPR43 in leptin secretory responses of murine adipocytes to short chain fatty acids. FEBS Lett 2010; 584:2381-6; PMID: 20399779; http://dx.doi.org/ 10.1016/j.febslet.2010.04.027.

[43] Perry RJ, Peng L, Barry NA, Cline GW, Zhang D, Cardone RL, Petersen KF, Kibbey RG, Goodman AL, Shulman GI. Acetate mediates a microbiome-brain-β- cell axis to promote metabolic syndrome. Nature, 2016, doi:10.1038/nature18309.

[44] Tuncil, Y.E., Thakkar, R.D., Marcia, A.D.R., Hamaker, B.R. and Lindemann, S.R., 2018. Divergent short-chain fatty acid production and succession of colonic microbiota arise in fermentation of variously-sized wheat bran fractions. Scientific reports, 8(1), pp.1-13.

[45] Qin, J., Li, R., Raes, J., Arumugam, M., Burgdorf, K.S., Manichanh, C., Nielsen, T., Pons, N., Levenez, F., Yamada, T. and Mende, D.R., 2010. A human gut microbial gene catalogue established by metagenomic sequencing. Nature, 464(7285), pp.59-65.

[46] De Filippis, F., Pellegrini, N., Vannini, L., Jeffery, I.B., La Storia, A., Laghi, L., Serrazanetti, D.I., Di Cagno, R., Ferrocino, I., Lazzi, C. and Turroni, S., 2016. High-level adherence to a Mediterranean diet beneficially impacts the gut microbiota and associated metabolome. Gut, 65(11), pp.1812-1821.

[47] Sonnenburg, E.D., Smits, S.A., Tikhonov, M., Higginbottom, S.K., Wingreen, N.S. and Sonnenburg, J.L., 2016. Diet-induced extinctions in the gut microbiota compound over generations. Nature, 529(7585), pp.212-215.

[48] Duncan, S.H., Russell, W.R., Quartieri, A., Rossi, M., Parkhill, J., Walker, A.W. and Flint, H.J., 2016. Wheat bran promotes enrichment within the human colonic microbiota of butyrate-producing bacteria that release ferulic acid. Environmental microbiology, 18(7), pp.2214-2225.

[49] Machiels K, Joossens M, Sabino J, De Preter V, Arijs I, Eeckhaut V, Ballet V, Claes K, Van Immerseel F, Verbeke K, Ferrante M. A decrease of the butyrate-producing species Roseburia hominis and Faecalibacterium prausnitzii defines dysbiosis in patients with ulcerative colitis. Gut. 2014;63(8): 1275-1283.

[50] Kim HJ, Bae SC. Histone deacetylase inhibitors: molecular mechanisms of action and clinical trials as anti-cancer drugs. American journal of translational research. 2011; 3(2):166.

[51] MacDonald VE, Howe LJ. Histone acetylation: where to go and how to get there. Epigenetics. 2009 ;4(3):139-143.

[52] Tolhurst G, Heffron H, Lam YS, Parker HE, Habib AM, Diakogiannaki E, Cameron J, Grosse J, Reimann F, Gribble FM. Short-chain fatty acids stimulate glucagon-like peptide-1 secretion via the G-protein–coupled receptor FFAR2. Diabetes. 2012;61(2):364-371.

[53] Tazoe H, Otomo Y, Karaki SI, Kato I, Fukami Y, Terasaki M, Kuwahara A. Expression of short-chain fatty acid receptor GPR41 in the human colon. Biomedical research. 2009;30(3): 149-156.

[54] Kendrick SF, O'Boyle G, Mann J, Zeybel M, Palmer J, Jones DE, Day CP. Acetate, the key modulator of inflammatory responses in acute alcoholic hepatitis. Hepatology. 2010;51(6):1988-1997.

[55] Usami M, Kishimoto K, Ohata A, Miyoshi M, Aoyama M, Fueda Y, Kotani J. Butyrate and trichostatin A attenuate nuclear factor κB activation and tumor necrosis factor α secretion and increase prostaglandin E2 secretion in human peripheral blood mononuclear cells. Nutrition research. 2008;28(5): 321-328.

[56] Xiong Y, Miyamoto N, Shibata K, Valasek MA, Motoike T, Kedzierski RM, Yanagisawa M. Short-chain fatty acids stimulate leptin production in adipocytes through the G protein-coupled receptor GPR41. Proceedings of the National Academy of Sciences. 2004;101(4):1045-1050.

[57] Kang I, Kim SW, Youn JH. Effects of nicotinic acid on gene expression: potential mechanisms and implications for wanted and unwanted effects of the lipid-lowering drug. The Journal of Clinical Endocrinology & Metabolism. 2011 Oct 1;96(10):3048-3055.

[58] Shimizu H, Masujima Y, Ushiroda C, Mizushima R, Taira S, Ohue-Kitano R, Kimura I. Dietary short-chain fatty acid

intake improves the hepatic metabolic condition via FFAR3. Scientific reports. 2019;9(1):1-0.

[59] Gao, X., Lin, S.H., Ren, F., Li, J.T., Chen, J.J., Yao, C.B., Yang, H.B., Jiang, S.X., Yan, G.Q., Wang, D. and Wang, Y., 2016. Acetate functions as an epigenetic metabolite to promote lipid synthesis under hypoxia. Nature communications, 7(1), pp.1-14.

[60] Martin AM, Lumsden AL, Young RL, Jessup CF, Spencer NJ, Keating DJ. The nutrient-sensing repertoires of mouse enterochromaffin cells differ between duodenum and colon. Neurogastroenterology & Motility. 2017;29(6):e13046.

[61] Dumont Y, Fournier A, St-Pierre S, Quirion R. Characterization of neuropeptide Y binding sites in rat brain membrane preparations using [125I][Leu31, Pro34] peptide YY and [125I] peptide YY3-36 as selective Y1 and Y2 radioligands. Journal of Pharmacology and Experimental Therapeutics. 1995; 272(2):673-680.

[62] Loh K, Herzog H, Shi YC. Regulation of energy homeostasis by the NPY system. Trends in Endocrinology & Metabolism. 2015; 26(3):125-135.

[63] Lin HV, Frassetto A, Kowalik Jr EJ, Nawrocki AR, Lu MM, Kosinski JR, Hubert JA, Szeto D, Yao X, Forrest G, Marsh DJ. Butyrate and propionate protect against diet-induced obesity and regulate gut hormones via free fatty acid receptor 3-independent mechanisms. PloS one. 2012;7(4).

[64] Grøndahl MF, Keating DJ, Vilsbøll T, Knop FK. Current Therapies That Modify Glucagon Secretion: What Is the Therapeutic Effect of Such Modifications?. Current diabetes reports. 2017;17(12):128.

[65] Holst JJ. The physiology of glucagon-like peptide 1. Physiological reviews. 2007 ;87(4):1409-1439.

[66] Sileikiene V, Mosenthin R, Bauer E, Piepho HP, Tafaj M, Kruszewska D, Weström B, Erlanson-Albertsson C, Pierzynowski SG. Effect of ileal infusion of short-chain fatty acids on pancreatic prandial secretion and gastrointestinal hormones in pigs. Pancreas. 2008;37 (2):196-202.

[67] Tang C, Ahmed K, Gille A, Lu S, Gröne HJ, Tunaru S, Offermanns S. Loss of FFA2 and FFA3 increases insulin secretion and improves glucose tolerance in type 2 diabetes. Nature medicine. 2015;21(2):173.

[68] Kimura I, Ozawa K, Inoue D, Imamura T, Kimura K, Maeda T, Terasawa K, Kashihara D, Hirano K, Tani T, Takahashi T. The gut microbiota suppresses insulin-mediated fat accumulation via the short-chain fatty acid receptor GPR43. Nature communications. 2013 May 7;4(1):1-2. PMID: 23652017; http://dx.doi.org/10.1038/ncomms2852.

[69] Zhao, L., Zhang, F., Ding, X., Wu, G., Lam, Y.Y., Wang, X., Fu, H., Xue, X., Lu, C., Ma, J. and Yu, L., 2018. Gut bacteria selectively promoted by dietary fibers alleviate type 2 diabetes. Science, 359(6380), pp.1151-1156.

[70] David LA, Maurice CF, Carmody RN, Gootengerg DB, Button JE, Wolfe B. & Turnbaugh, PJ (2013). Diet rapidly and reproducibly alters the human gut microbiome. Nature.;505:7484.

[71] Kumar J, Rani K, Datt C. Molecular link between dietary fibre, gut microbiota and health. Molecular Biology Reports. 2020; 4:1-9.